Produced Water Treatment
Field Manual

Produced Water
Treatment Field Manual

Maurice Stewart
Ken Arnold

AMSTERDAM • BOSTON • HEIDELBERG • LONDON • NEW YORK • OXFORD
PARIS • SAN DIEGO • SAN FRANCISCO • SINGAPORE • SYDNEY • TOKYO
Gulf Professional Publishing is an imprint of Elsevier

Gulf Professional Publishing is an imprint of Elsevier
225 Wyman Street, Waltham, MA 02451, USA
The Boulevard, Langford Lane, Kidlington, Oxford, OX5 1GB, UK

Library of Congress Cataloging-in-Publication Data
Application Submitted.

British Library Cataloguing-in-Publication Data
A catalogue record for this book is available from the British Library.

ISBN: 978-1-85617-984-3

For information on all Gulf Professional Publishing
publications visit our Web site at www.elsevierdirect.com

10 11 12   10 9 8 7 6 5 4 3 2 1

Printed and bound in the USA

Contents

Part 1
Produced Water Treating Systems

Contents

INTRODUCTION

When hydrocarbons are produced, the well stream typically contains water produced in association with these hydrocarbons.

Water must be separated from the hydrocarbons and disposed of in a manner that does not violate established environmental regulations.

Produced water is separated from hydrocarbons by passing the well stream through gravity separation process equipment such as:

> Three-phase separators
>
> Heater treaters, and/or
>
> Free-water knockout vessel

DOI: 10.1016/B978-1-85617-984-3.00001-8

Gravity separation equipment is characterized by the following:

Cannot achieve a full 100% separation of hydrocarbons from the produced water

Produced water will contain 0.1 to 10 volume percent of dispersed and dissolved hydrocarbons

Produced water treating facilities are used to reduce the hydrocarbon content in the produced water prior to final disposal.

Regulatory standards for overboard disposal of produced water into offshore surface waters vary from country to country.

Failure to comply with regulations can result in:

Civil penalties

Large fines

Lost or deferred production

Currently, regulations require the "total oil and grease" content of the effluent water to be reduced to levels ranging from between 15 mg/l and 50 mg/l.

Disposal or produced water into onshore surface waters is generally prohibited by environmental regulations.

Onshore disposal typically requires the produced water effluent to be injected into a saltwater disposal well.

The purpose of this section is to present to the designer with a procedure for:

Selecting the appropriate type of equipment for treating oil from produced water, and

Provide the theoretical equations and empirical rules necessary to size the equipment

When this design procedure is followed, the designer will be able to:

Develop process flow diagram

Determine equipment sizes

Evaluate vendor proposals for any wastewater treating system once the following parameters are determined:

Discharge quality

Produced water flow rate

Oil specific gravity

Water specific gravity

Drainage requirements

DISPOSAL STANDARDS

Offshore Operations

Standards for the disposal of produced water are developed by the government regulatory agencies.

Regulatory agencies generally specify the analytical for determining the oil content.

Produced water toxicity is regulated only in the United States where a government permit is necessary to limit toxicity of produced water discharged into the waters.

Table 1-1 summarizes offshore disposal standards for several countries. The standards are current as of this writing.

Onshore Operations

Disposal of produced water into freshwater streams and rivers is generally prohibited except for the very limited cases where the effluent is low in salinity.

Some oilfield brines might kill freshwater fish and vegetation due to high salt content.

Regulatory agencies generally:

Require produced water from onshore operations be disposed of by subsurface injection

Regulate the completion and operation of the disposal wells

Table 1-1 Worldwide Produced Water Effluent Oil Concentration Limitations

Country	Oil Concentration/Limitations
Ecuador, Columbia, Brazil	30 mg/l All facilities
Argentina and Venezuela	15 mg/l New facilities
Indonesia	25 mg/l All facilities Zero discharge inland water
Malaysia, Middle East	30 mg/l All facilities
Nigeria, Angola, Cameroon, Ivory Coast	50 mg/l All facilities
North Sea, Australia	30 mg/l All facilities
Thailand, Brunei	30 mg/l All facilities
U.S.A.	29 mg/l OCS waters Zero discharge inland water

CHARACTERISTICS OF PRODUCED WATER

Produced water contains a number of substances, in addition to hydrocarbons, that affect the manner in which water is handled.

Composition and concentration of substances may:

Vary between fields

Vary between different production zones within a single field

Concentration is expressed in milligrams per liter (mg/l), which is mass per volume ratio and is approximately equal to parts per million (ppm).

Some of the important produced water constituents are discussed in this section.

Dissolved Solids

Produced waters contain dissolved solids, but the amount varies from less than 100 mg/l to over 300,000 mg/l, depending on:

Geographical location

Age and type of reservoir

Water produced with gas is condensed water vapor with few dissolved solids and will be fresh with very low salinity.

Aquifer water produced with gas or oil will be much higher in dissolved solids.

Produced water from hot reservoirs tend to have higher TDS concentrations while cooler reservoirs tend to have lower levels of TDS.

Dissolved solids are inorganic constituents that are predominantly sodium (Na^+) cations and chloride (Cl^-) anions.

Other common cations are:

Calcium (Ca^{2+})

Magnesium (Mg^{2+})

Iron (Fe^{2+})

Cations encountered less frequently are:

Barium (Ba^{2+})

Potassium (K^+)

Stontium (Sr^+)

Aluminum (Al^{3+})

Lithium (Li^+)

Other antions present are:

Bicarbonate (HCO_3)

Carbonate (CO_3)

Sulfate (SO_4)

All produced water treating facilities should have water analysis data for:

Each major reservoir, and

Each combined produced water stream

Particularly important are constituents that could precipitate to form scales.

Precipitated Solids (Scales)

The more troublesome ions are those that react to form precipitates when pressure, temperature, or composition changes.

They form deposits that form in tubing, flowlines, vessels, and produced water treating equipment.

Mixing of oxygenated deck drain water with produced water should be avoided because this may result in the formation of calcium carbonate ($CaCO_2$), calcium sulfate ($CaSO_4$), and iron sulfide (FeS_2) scale, along with oil-coated solids.

SCALE REMOVAL

Hydrochloric acid can be used to dissolve calcium carbonate and iron sulfide scales.

However, iron sulfide chemically reacts with hydrochloric acid and produces hydrogen sulfide.

Calcium sulfate is not soluble in hydrochloric acid, but chemicals are available that will convert it to an acid-soluble form that can than be removed by the acid.

This two-step process is slow.

Removal of calcium sulfate is more difficult than the removal of calcium carbonate.

Practical means of dissolving barium or strontium sulfate are not available.

These hard scales are removed by mechanical means which is time-consuming.

Mechanical removal of scale creates a disposal problem for the waste material and the possibility of forming NORM.

CONTROLLING SCALE USING CHEMICAL INHIBITORS

Scale-inhibiting chemicals are available to retard or prevent all types of scale.

They mostly function by enveloping a newly precipitated crystal, thereby retarding growth.

Common scale inhibitors include:

Inorganic phosphates (inexpensive but only applicable at low temperature)

Organic phosphate esters (easy to monitor but limited to temperatures below 100°F (38°C)

Phosphates (easy to monitor and have a higher thermal stability to 150°F (57°C)

Polymers (best thermal stability and effectiveness, but difficult to monitor)

SAND AND OTHER SUSPENDED SOLIDS

Produced water often contains other suspended solids including:

Formation sands and clays

Stimulation (fracturing) proppant

Miscellaneous corrosion products

Amount of suspended solids is usually small unless the well is producing from a unconsolidated formation, in which case large volumes of sand can be produced.

Produced sand is characterized by the following:

Usually oil wet

Disposal is a problem

Sand removal is discussed in the later section "Equipment Description and Sizing."

Small amounts of solids in produced water may or may not create problems in water treating depending on:

> Particle size

> Its relative attraction to the dispersed oil

If the physical characteristics and electronic charge of such solids result in an attraction to the dispersed oil droplets, the solid particles:

> Can attach to the dispersed oil droplets to stabilize emulsions

>> This prevents coalescence and separation of the oil phase

The combined specific gravity of the resulting oil/solid droplet can be approximately equal to that of the produced water, and gravity separation becomes difficult, if not impossible.

The concentration of suspended solids can be monitored with 0.45-micron Millipore filter test, and residue is analyzed for mineral content in an attempt to identify the source of the solid.

When solids are present, the following practices should be applied:

> Chemical treatment must be used to "break" the electronic attraction between the solid and the oil droplet.

> Equipment design must incorporate solids removal ports, jets, and/or plates.

> Oil measurement techniques not affected by the solids should be used.

> Solids are likely to be oil-coated, and offshore disposal may be prohibited.

>> This is the case in the United States.

>> Applies to solids removed from desanders or vessels, not to the solids suspended in the water.

Water injection for disposal should be made into a disposal zone that has high enough pore space openings to prevent the suspended solids from plugging.

> Consideration should be given to using filtration equipment to remove the larger particles prior to injection into the disposal well.

> Periodic back flowing and acidizing are generally needed to maintain disposal wells if filtration is not applied.

Water-flood injection for pressure maintenance and additional recovery often requires filtration (to remove suspended solids).

Water injection pressures typically must be maintained below the fracture pressure of the formation.

Dissolved Gases

The most important gases found in produced water include:

Natural gas (methane, ethane, propane and butane)

Hydrogen sulfide

Carbon dioxide

In the reservoir the water can be saturated with these gases at relatively high pressures.

As the produced water flows up the wells:

Most of these gases flash to the vapor phase

Most of these gases are removed in primary separators and stock tanks

The pressures and temperatures at which the produced water is separated from the main oil, condensate, and/or natural gas streams:

Impact the quality of dissolved gas that will be contained in the produced water stream feeding the water treating facilities.

The higher the separation pressure, the higher the quantity of dissolved gases will be.

The higher the separation temperature, the lower the quantity of dissolved gases will be.

Natural gas has the following characteristics:

Components are slightly soluble in water at moderate to high pressures and will be present in the produced water stream.

Components have an affinity for dispersed oil droplets, and this principle is applied to the design of gas floatation equipment used in produced water treating systems.

Hydrogen sulfide:

Will be present in the produced water stream if:

Hydrogen sulfide is present in the produced reservoir fluid, or

Sulfate-reducing bacteria are a problem in the reservoir or production equipment

Is corrosive

Can cause iron sulfide scaling, and

Is extremely toxic if inhaled.

The toxicity of hydrogen sulfide:

Hinders operation and maintenance of equipment, particularly when the vessels must be opened for adjustments, as in the case when weir adjustments are required in gas flotation cells

Requires personnel to take specialized training and wear life support breathing equipment

Iron sulfide (the corrosive product of hydrogen sulfide) presents a potential fire hazard since it is prone to auto-ignition when exposed to air or other sources of oxygen.

Carbon dioxide :

Will be present in the produced water if CO_2 is present in the produced reservoir fluid

Is corrosive and can cause $CaCO_3$ scaling

Removal of H_2S and CO_2 will result in increased pH, which could lead to scaling

Oxygen is characterized by the following:

It is not found naturally in produced water

When produced water is brought to the surface and exposed to the atmosphere, oxygen will be absorbed into the water

Water containing dissolved oxygen can cause

Severe and rapid corrosion

Solids generation from oxidation reactions

Oil weathering, which inhibits cleanup

To prevent the above, a natural gas blanket should be maintained on all of the production and water treating tanks and vessels used within the process

Seawater is characterized by the following:

It is often used as the source of water for water floods and water injection pressure maintenance projects offshore.

It contains considerable amounts of dissolved oxygen and some carbon dioxide.

It may also contain bacteria.

Oxygen and carbon dioxide must be removed from the source water by either vacuum de-aeration or gas stripping prior to injection.

Oil-in-Water Emulsions

"Normal emulsions" are encountered in most oil fields.

Water droplets are dispersed in the oil continuous phase.

Water is dispersed in the form of small droplets ranging between 100 and 400 microns in diameter.

"Reverse emulsions" can occur in produced water treating operations.

Oil droplets are dispersed in the water continuous phase.

Oil is dispersed in the form of very small droplets typically less than 150 microns in diameter.

Unstable emulsions

Oil droplets will coalesce when they come in contact with each other and form larger droplets, thus breaking the emulsion.

Unstable emulsions usually break within minutes and do not require any treatment

Stable emulsions

Stable emulsions are the suspension of two immiscible liquids in the presence of a stabilizer or emulsifying agent that act to maintain an interfacial film between the phases.

Chemicals, heat, settling time, and electrostatics are used to alter and remove the film and cause emulsion breakdown.

Untreated, stable emulsions can remain for days or even weeks.

Water-in-Oil Emulsions

Emulsion breakers are known as "destabilizers" or "demulsifiers." They are:

Oil-soluble

Added to the total well stream ahead of the process equipment

Being oil-soluble:

> The emulsion breaker is carried with the crude.
>
> Thus, if the emulsion is not broken in the first-stage separator, the chemical has additional time to act in the subsequent separators and the stock tank.

Oil-in-Water Emulsions

> Emulsion breakers are known as "reverse emulsion breakers" which are special destabilizers or demulsifiers. They are:
>
> > Similar to conventional emulsion breakers except they are water-soluble
> >
> > Generally injected into the water stream after the first oil-in-water separation vessel
> >
> > > Typical concentrations are in the 5–15 ppm range.
> >
> > Overtreating should be avoided because these chemicals can stabilize an emulsion.

Emulsions in produced water are characterized by the following:

> They will become oil in the form of dispersed droplets after the emulsion film is broken.
>
> Droplets will coalesce to yield an oily film that can be separated from the produced water by using gravity settling devices such as skim vessels, coalescers, and plate separators.
>
> Small droplets require excessive gravity settling time, so flotation cells or acceleration enhanced methods such as hydrocyclones and centrifuges are used.
>
> Equipment selection is based on the inlet oil's droplet diameter and concentration.

Dissolved Oil Concentrations

Dissolved oil:

> Is also called "soluble oil"
>
> Represents all hydrocarbons and other organic compounds that have some solubility in produced water

The source of the produced water affects the quantity of the dissolved oil present.

Produced water derived from gas/condensate production typically exhibits higher levels of dissolved oil.

Process water condensed from glycol regeneration vapor recovery systems contains aromatics including benzene, toluene, ethyl benzene, and xylenes (BTEX) that are partially soluble in produced water.

Gravitational-type separation equipment will not remove dissolved oil.

Thus, a high level of total oil and grease could be discharged if the produced water source contains significant quantities of dissolved oil.

Produced water streams containing high concentrations of dissolved oil can be recycled to a fuel separator to help reduce the quantity of dissolved oil in the water.

Other technologies are currently being evaluated but such processes are not yet readily available for commercial applications. Technologies include:

Bio-treatment

Adsorption filtration

Solvent extraction

Membranes

Lab Tests

It is essential that actual water test analysis data for dissolved and dispersed oil concentrations are needed in the planning stage prior to designing a water-treating facility for a specific application.

If the design engineer assumes a value for the dissolved oil content without first having obtained actual water test analysis for the specific produced water stream to be treated, the facility design may not be capable of treating the effluent water to meet regulatory requirements.

Dispersed Oil

Oil droplets range from 0.5 to 200 microns in diameter.

"Oil size droplet distribution" is important for the following reasons:

It is one of the key parameters influencing the produced water treating performance.

According to Stokes' Law:

The rising velocity of an oil droplet is proportional to the square of the droplet diameter.

For equipment that operates on the principle of Stokes' law, the diameter of the oil droplet has a major effect on the separation and removal of the oil droplet from the water.

The capability of a given de-oiling device or system to remove and recover dispersed oil decreases as the droplet size decreases.

Oil droplet size distribution is a fundamental characteristic of produced water and must be considered in designing and sizing treating systems to meet regulatory standards for effluent water compliance.

The histogram of an oil droplet distribution (Figure 1-1) shows the following:

The histogram divides the particle counts into discrete size ranges along the horizontal axis.

The number and size of the ranges are determined by the equipment used to obtain the data, for example, a Coulter counter.

The height of the vertical bars corresponds to the volume percentage of oil droplets in each range.

A particle distribution curve is constructed by connecting the tops of the bars at the midpoint of each size range.

An oil volume distribution curve (Figure 1-2) shows the following:

The volume percentage of the particles is equal to or smaller than each specified size that is plotted.

The vertical axis scale is from 0 to 100% since the data are plotted cumulatively.

Oil droplet size distribution in a produced water system may vary from:

Point to point, and

From one system to another

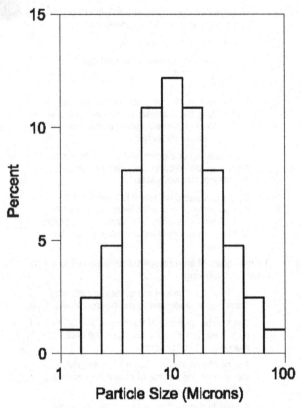

FIGURE 1-1 Histogram of oil droplet distribution.

Size distribution is affected by:

Interfacial tension

Turbulence

Temperature

System shearing (pumping, pressure drop across pipe fitting, etc.)

Other factors

Dispersed oil should be measured in the field when troubleshooting and/or upgrading systems, whenever possible.

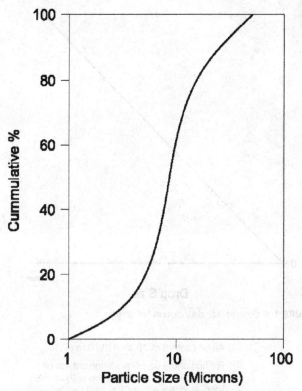

FIGURE 1-2 Typical oil volume distribution curve.

In the absence of data, the generalized relationship in Figure 1-3 can be used for oil droplet size distributions.

Since the distribution is linear, it places more volume in smaller-diameter droplets.

Since this relationship is a very rough estimate, field data should be used whenever possible.

In the absence of field data, the following can be used:

For produced water effluent from a three-phase separator, the following data can be used:

A maximum oil droplet diameter of 250 to 500 microns

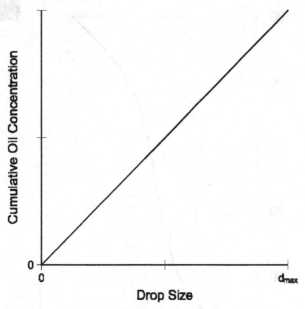

FIGURE 1-3 Droplet size distribution for design.

An oil content of 1000 to 2000 mg/l

For first phase de-oiling equipment, an oil droplet diameter of 30 microns with an inlet total oil level less than 100 mg/l can be assumed for produced water feed to final treating equipment.

Operational experience in the area may also provide reliable data from similar facilities that can be used to estimate inlet oil concentrations and droplet size distributions.

SYSTEM DESCRIPTION

Table 1-2 lists the various methods employed in produced water treating systems and the types of equipment that employ each method.

Figure 1-4 shows a typical produced water system configuration.

Table 1.2 Produced Water Treating Equipment

Approximate Minimum Method	Equipment Type	Drop Size Removal Capacities (Microns)
Gravity separation	Skimmer tanks and vessels	
	API separators	100–150
	Disposal piles	
	Skim piles	
Plate coalescence	Parallel plate interceptors	
	Corrugated plate interceptors	30–50
	Cross-flow separations	
	Mixed-flow separators	
Enhanced coalescence	Precipitators	
	Filters/coalesces	10–15
	Free-flow turbulent coalesces	
Gas flotation	Dissolved gas	
	Hydraulic dispersed gas	10–20
	Mechanical dispersed gas	
Enhanced gravity Separation	Hydrocyclones	10–30
	Centrifuges	
Filtration	Multimedia membrane	1+

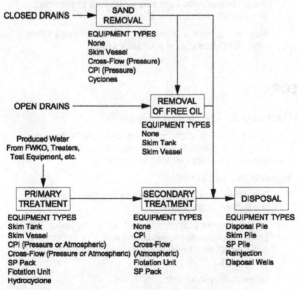

FIGURE 1-4 Typical produced water treating system.

Produced water will always have some form of primary treating prior to disposal.

Depending upon the severity of the treating problem, secondary treatment may be required.

Offshore produced water :

>   Can be piped directly overboard after treating, or

>   Can be routed through a disposal pile or a skim pile

Water from the deck drains must be treated for removal of "free" oil.

>   Normally done in a skim vessel called a sump tank

>   Water from the sump tank is either combined with the produced water or routed separately for disposal overboard

Onshore produced water:

>   Normally reinjected in the formation, or

>   Pumped into a disposal well

Closed drains

>   Should never be tied into atmospheric drains

>   Should be routed to a pressure vessel prior to entering an atmospheric skim tank or pile

>   Should be done in a skim vessel, with or without a CPI or cross-flow separator, in a pressure vessel

THEORY

Function of all water-treating equipment :

>   Cause the oil droplets, which are dispersed in the water continuous phase, to separate and float to the surface of the water so they can then be removed

In gravity separation units:

>   The difference in specific gravity causes the oil to float to the surface of the water.

>   Oil droplets are subjected to continuous dispersion and coalescence during the trip up the well bore through the surface chokes, flowlines, control valves, and process equipment.

>   When energy input rate:

>>   Is high, the drops are dispersed to smaller sizes

Is low, small droplets collide and join together in the coalescence process

The three basic phenomena that are used in the design of common produced water treating equipment are:

Gravity separation

Coalescence

Flotation

Dispersion also affects the design but to an unpredictable degree.

In the past, filtration has been tried but, due to high maintenance costs, has been found to be unsatisfactory.

Gravity Separation

Most commonly used water treating equipment depends on the forces of gravity to separate the oil droplets from the water continuous phase.

Oil droplets, being lighter than the volume of water they displace, have a buoyant force exerted upon them.

This force is resisted by a drag force caused by their vertical movement through the water.

When the two forces are equal, a constant velocity is reached, which can be computed from Stokes' Law as:

Field units

$$V_r = \frac{1.78 \times 10^{-6}(\Delta SG)d_0^2}{\mu_w} \qquad (1\text{-}1a)$$

SI units

$$V_r = \frac{5.6 \times 10^{-7}(\Delta SG)d_0^2}{\mu_w} \qquad (1\text{-}1b)$$

where

V_r = rising velocity of the oil droplet, ft./sec (m/sec)

d_0 = diameter of oil droplet, microns (μ)

ΔSG = difference in specific gravity between oil and water, relative to water

μ_w = viscosity of the continuous water phase, cp

Conclusions drawn from Stokes' Law are:

The larger the size of an oil droplet, the larger the square of its diameter and, thus, the greater its vertical velocity will be.

The greater the difference in density between the oil droplet and the water phase, the greater the vertical velocity will be.

The higher the temperature, the lower the viscosity of the water and, thus, the greater the vertical velocity will be.

Coalescence

The process of coalescence in water-treating systems is more time dependent than the process of dispersion.

In a dispersion of two immiscible liquids, immediate coalescence seldom occurs when two droplets collide.

If the droplet pair is exposed to turbulent pressure fluctuations, and kinetic energy of the oscillations induced in the droplet pair is larger than the energy of adhesion between them, the contact will be broken before coalescence is completed.

If the energy input into the system is too great, dispersion will occur, as discussed below.

If there is no energy input, then the frequency of droplet collision, which is necessary to indicate coalescence, will be low, and coalescence will occur at a very low pace.

Most water treating equipment, with the exception of flotation units and hydrocyclones, consists of vessels in which the oil droplets rise to a surface due to gravity forces.

From a process standpoint, these are considered "deep bed gravity settlers."

Experiments with deep bed gravity settlers yield the following two qualitative conclusions:

Doubling the residence time causes only a 10% increase in the maximum size droplet that will be grown in a gravity settler.

The more dilute the dispersed phase (oil), the greater the residence time required to grow a given particle size. That is, coalescence occurs more rapidly in concentrated dispersions.

From these conclusions it shows that after an initial period of coalescence in a settler, additional retention time has a rapidly diminishing ability to cause coalescence and to capture oil droplets.

Dispersion

Refers to the process of a discontinuous phase (oil) being split into small droplets and distributed throughout a continuous phase (water).

This dispersion process occurs when a large amount of energy is input to the system in a short period of time.

This energy input overcomes the natural tendency of two immiscible fluids to minimize the contacting surface area between the two fluids.

The dispersion process is diametrically opposed by coalescence, which is the process in which small droplets collide and combine into larger droplets.

As the oil and water mixture flows through the piping, these two processes are simultaneously occurring.

In the piping a droplet of oil splits into smaller droplets when kinetic energy of its motion is larger than the difference in surface energy between the single droplet and the two smaller droplets formed from it.

While this process is occurring, the motion of the smaller oil droplets causes coalescence to occur.

Therefore, it should be possible to define statistically a maximum droplet size for a given energy input per unit mass and time at which the rate of coalescence equals the rate of dispersion.

One relationship for the maximum particle size that can exist at equilibrium was proposed by Hinze as follows:

$$d_{max} = 432(t_r/\Delta P)2/5(\sigma/\rho w)3/5 \qquad (1\text{-}2)$$

where

D_{max} = diameter of droplet above whose size only by 5% of the oil volume is contained, microns

σ = surface tension, dyne/cm

ρ_w = density, g/cm^3

ΔP = pressure drop, psi

t_r = retention time, min

From Equation 1-2, it can be seen that the greater the pressure drop and, thus, the shear forces that the fluid experiences in a given period of time while flowing through the treating system, the smaller the maximum oil droplet diameter will be.

Large pressure drops that occur in small distances through chokes, orifices, throttling globe control valves, desanders, and so on, result in smaller droplets.

Equation 1-2 can be applied to determine a maximum droplet size that can exist downstream of a control valve or any other device that causes a large pressure drop.

The dispersion process is theoretically not instantaneous.

However, it appears from field experience to occur very rapidly.

For design purposes it could be assumed that whenever large pressure drops occur, all droplets larger than D_{max} will instantaneously disperse.

Unfortunately, Equation 1-2 cannot be used directly to predict the coalescence of droplets that occur in piping with high-pressure drops downstream of a process component in which dispersion takes place.

This is because the coalescence to a new D_{max} determined in Equation 1-2 is time-dependent, and there is currently no basis to estimate the time required to grow D_{max}.

Flotation

The flotation process is characterized by the following:

Involves the injection of fine gas bubbles into the water phase

Gas bubbles in the water adhere to the oil droplets

Buoyant force on the oil droplet is greatly increased by the presence of the gas bubble

Oil droplets are then removed when they rise to the water surface, where they are trapped in the resulting foam and skimmed off the surface

Experimental results show:

Very small oil droplets (greater than 10 microns) in a very dilute suspension can be removed.

High percentages (90% +) of oil removal are achieved in very short times.

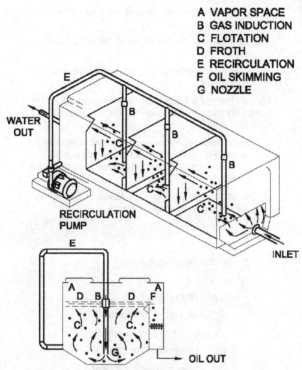

A VAPOR SPACE
B GAS INDUCTION
C FLOTATION
D FROTH
E RECIRCULATION
F OIL SKIMMING
G NOZZLE

CROSS SECTION THROUGH CELL

FIGURE 1-5 Dispersed gas flotation unit with inductor.

Figure 1-5 shows a cross section of a one-cell and a three-cell hydraulic inductor dispersed gas flotation unit.

Clean water from the effluent is pumped to a recirculation header (E) that feeds a series of venture educators (B).

Water flowing through the educator sucks gas from the vapor space (A) that is released at the nozzle (G) as a jet of small bubbles.

The bubbles rise, causing flotation in the chamber (C), forming a froth (D) that is skimmed with a mechanical device at (F).

It is extremely difficult to develop a precise mathematical model of the process occurring in the zones identified in this cross section.

However, with the aid of some liberal assumptions it is possible to:

Develop a qualitative model of the efficiency of such a cell

Gain an understanding of the importance of various parameters

The efficiency of a specific cell with constant geometry can be approximated through the use of Equations 1-3 through 1-5.

Since these equations are presented to provide a qualitative "feel" for the effects of various parameters on flotation cell efficiency, units are not listed.

In using these equations, however, one must use parameters with consistent units.

$$E = (Ci - Co)/Ci \qquad (1\text{-}3)$$

$$E = K/(Qw - K') \qquad (1\text{-}4)$$

$$E = (6\pi K_p r^2 h q_g)/(q_w d_b) \qquad (1\text{-}5)$$

where

E = efficiency per cell

C_i = inlet oil concentration

C_o = outlet oil concentration

Q_w = liquid flow rate, BPD

K_p = mass transfer coefficient

r = radius of mixing zone

h = height of mixing zone

q_g = gas flow rate

q_w = liquid flow through the mixing zone

d_b = diameter of gas bubble

The following conclusions can be drawn from Equation 1-5:

Removal efficiency is independent of the influent oil concentration or the oil droplet size distribution.

Decreasing the diameter of the gas bubbles without changing the gas flow rate increases the efficiency.

Increasing the gas flow rate increases the efficiency.

Increasing the bulk flow rate decreases the efficiency.

Equation 1-5 cannot be used directly as:

It depends on the design details of the particular unit, which is under the control of the manufacturer

It depends on the mass flow transfer coefficient, which is a function of the composition and chemical treatment of the liquid

Most manufactures attempt to design each cell for a typical efficiency in excess of 50%.

The overall efficiency of a multiple cell flotation unit can be calculated from Equation 1-6:

$$E_t = 1 - [1 - E]^n \qquad (1\text{-}6)$$

where

E_t = overall efficiency

N = number of stages or cells

For an average design efficiency of 50% per stage, the following overall efficiencies may be calculated:

Overall # of Cells	Efficiency (E_t)
1	0.50
2	0.75
3	0.87
4	0.94
5	0.97

Number of Cells

Most flotation cells consist of three or four cells.

Using more cells may not be cost-effective for the small performance increases shown above.

Retention time:

Each cell must have some retention time so that the gas bubbles may have time to rise to the liquid surface.

It is recommended that a minimum water retention time of one minute be provided in each cell.

Flow considerations:

Units function best if the water flow through the unit is smooth.

It is recommended that throttling level controls be used to control the level in the upstream components of the system and in the flotation unit.

Filtration

The flow of produced water through a properly selected filter media will cause the small droplets of oil to contact and attach to the filter fibers.

Depending on the media design and thickness, these droplets will either stay trapped in the media or eventually "grow" as other droplets contact them.

At some point the droplets will become large enough so that the drag forces on the droplet created by the bulk water flow through the media cause the now larger droplets to be stripped from the media.

These larger droplets are then more easily separated by gravity settling downstream of the media. This action is called "filter/coalescing."

It is also possible to design the filter media to drop the oil.

The media is cleaned periodically by stopping the flow and backwashing that is, flowing at very high velocities in the reverse direction for a short period of time.

Thus a standard filter such as those described in Part 2 can be used.

EQUIPMENT DESCRIPTION AND SIZING

Skim Tanks and Skim Vessels

General Considerations

The simplest form of primary equipment is a skim (clarifier) tank or vessel (Figure 1-6).

It is designed to provide long residence times during which coalescence and gravity separation can occur.

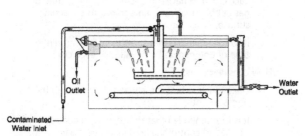

Contaminated
Water Inlet

FIGURE 1-6 Schematic of a skimmer tank.

Skim tanks can be used as:

Atmospheric tanks

Low-pressure vessels

Surge tanks ahead of other produced water treating equipment

The terminology is sometimes confusing.

Skim (clarifier) tanks

Tanks used to remove dispersed oil

Settling tanks

Tanks whose primary purpose is to remove entrained solids

Wash tanks

Function as a free-water knockout or gunbarrel

Used when the incoming stream contains 10 to 90% oil

Designed to make a rough separation of the oil and water

Water from wash tanks is sent to a skim (clarifier) tank or another unit to remove the remaining oil

If the desired outlet oil concentration (ppm) is known, the theoretical dimensions of the vessel can be determined.

Unlike separation, we cannot ignore the effects of vibration, turbulence, short circuiting, and so on.

API Publication 421, "Management of Water Discharges: Design and Operation of Oil-Water Separators," uses short-circuit factors as high as 1.75.

It is the basis upon which many of the sizing formulas were derived.

Configurations

Skim vessels can be either vertical or horizontal in configuration.

Vertical (Figure 1-7)

Oil droplets must rise upward countercurrent to the downward flow of the water.

Some skimmers have inlet spreaders and outlet collectors to help even the distribution of the flow.

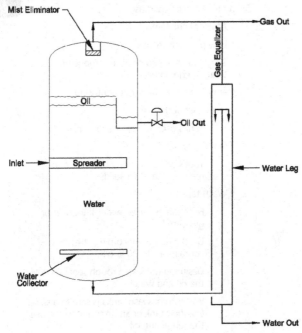

FIGURE 1-7 Schematic of a vertical skimmer vessel.

Oil, water, and any flash gases are introduced from the water help to "float" the oil droplets.

In the quiet zone between the spreader and the water collector, some coalescence can occur, and the buoyancy of the oil droplets causes them to rise counter to the water flow.

Oil will be collected and skimmed off the surface.

The thickness of the oil pad depends on the relative heights on the oil weir and the water leg and on the difference in specific gravity of the two liquids.

An interface level controller is often used in place of a water leg.

Horizontal (Figure 1-8)

Oil droplets rise perpendicular to the flow of the water.

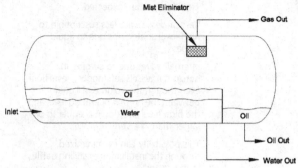

FIGURE 1-8 Schematic of a horizontal skimmer vessel.

The inlet enters in the water section so that the flashed gases may act as a dissolved gas flotation cell.

Water flows horizontally for most of the length of the vessel.

Baffles could be installed to straighten the flow.

Oil droplets coalesce in this section of the vessel and rise to the oil-water surface, where they are captured and eventually skimmed over the oil weir.

The height of the oil can be controlled by:

Interface control

Water leg similar to that shown in Figure 1-7, or

Bucket and weir arrangement

Horizontal vessels are more efficient at water treating because the oil droplets do not have to flow countercurrent to the water flow.

Vertical skimmers are used in instances where:

1.   Sand and other solid particles must be installed.

This can be done with either the water outlet or a sand drain off the bottom.

Experience with elaborately designed sand drains in large horizontal vessels is expensive, and they have been only marginally successful in field operations.

2.   Liquid surges are expected.

Vertical vessels are less susceptible to high-level shutdowns due to liquid surges.

Internal waves due to surging in horizontal vessels can trigger a level float even though the volume of liquid between the normal operating level and the high-level shutdown is equal to or larger than in a vertical vessel.

This possibility can be minimized through the installation of stilling baffles in the vessel.

Vertical vessels have some drawback that are not process-related and need to be considered when making a selection. Some examples are:

The PSV and some of the controls may be difficult to service without special access platforms and ladders.

The vessel may have to be removed from a skid for trucking due to height restrictions.

Pressure versus Atmospheric Vessels

The choice between a pressure or an atmospheric vessel is not determined solely by the water treating requirements.

The overall needs of the system must be considered.

Pressure vessels are more expensive than tanks. However, they are recommended where:

1.   Potential gas blow-by through the upstream vessel dump system could create too much back pressure in an atmospheric vent system.

2.   The water must be dumped to a higher level for further treating and a pump would be needed if an atmospheric vessel were installed.

Due to the danger from overpressure and potential gas venting problems associated with atmospheric vessels, pressure vessels are preferred downstream of pressurized three-phase separators.

An individual cost/benefit decision must be made for each application, taking into account all the requirements of the system.

Retention Time

Skim tanks are often used as the primary produced water treating equipment.

Oil concentration of the inlet water entering the skim tank ranges from 500 to 10,000 mg/l.

A minimum residence time of 10 to 30 minutes should be provided to

Assure that surges do not upset the system

Provide for some coalescence

The minimum droplet size removal is in the 100 to 300 micron range.

The potential benefits of providing much more residence time will probably not be cost-efficient beyond this point.

Skimmers with long residence times require baffles to attempt to distribute the flow and eliminate short-circuiting.

Tracer studies have shown that skimmer tanks, even those with carefully designed inlet spreaders and outlet baffles, exhibit short circuiting and poor flow behavior.

This result is probably due to density and temperature differences, deposition of solids, and corrosion of spreaders.

Figure 1-9 is a schematic of a vertical skim tank with baffles.

Performance Considerations

Several factors can affect the performance of a skim tank. Some of the more important factors include:

Carefully designed inlet and outlet distributors significantly improve the performance of a skim tank.

Higher inlet water temperatures improve the oil removal due to a reduction in the bulk water phase viscosity.

A short, wide, "stocky" tank is preferred over a tall, slender tank because it offers a lower downward water velocity, which aids in gravity separation.

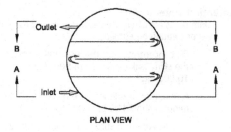

PLAN VIEW

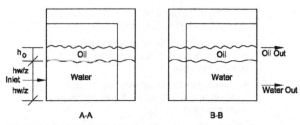

FIGURE 1-9 Schematic of a vertical skim tank with baffles.

Unbaffled tanks are inefficient due to short-circuiting.

Horizontal baffles improve skim tank performance, however, to achieve maximum benefit, they should be installed as close to the horizontal as possible and caution should be used during operating maintenance not to alter the baffle arrangement.

Often water-treating chemicals, such as flocculants, are added upstream of the skim vessel.

These work effectively to remove the smaller oil droplets by attaching to the oil droplets and causing them to rise to the oil–water interface in the skim vessel

If chemical dosage is not carefully monitored, especially when the water rate decreases, an excess of chemical flocculants will result in a froth layer at the oil-water interface.

This froth can cause the level controller to malfunction, leading to oil potentially spilling out of the vessel.

Skim vessels are recommended when:

Pressure reduction from a separator is required to protect downstream produced water treating equipment

Degassing water, catching oil slugs, or controlling surges is desired and the skim vessel is between the upstream separator and downstream produced water treating equipment

An existing vessel can be converted or space is available for a new vessel

The inlet oil concentration is high and the effluent must be reduced to 250 mg/l for the downstream equipment

Solid contaminants are in the inlet stream

Skim vessels are not recommended when:

Influent oil droplet sizes are mostly below 100 microns

Size and weight are the primary considerations

Offshore structure (platform, tension leg, etc.) movement could generate waves in the vessel

Water temperature is very cold due to long pipelines connected to other platforms

Skimmer Sizing Equations
Horizontal Cylindrical Vessels

The required diameter and length of a horizontal cylinder operating 50% full of water can be determined from Stokes' law using an efficiency factor of 1.8 for turbulence and short-circuiting:

Settling criteria

Field units

$$dL_{eff} = \frac{1000\beta_w Q_w \mu_w}{\alpha_w (\Delta SG)(d_m)^2} \qquad (1\text{-}7a)$$

SI units

$$dL_{eff} = \frac{1,145,734\beta_w Q_w \mu_w}{\alpha_w (\Delta SG)(d_m)^2} \qquad (1\text{-}7b)$$

where

d = vessel diameter, in. (mm)

Q_w = water flow rate, bwpd (m^3/hr)

μ_w = water viscosity, cp

d_m = oil droplet diameter, microns

L_{eff} = effective length in which separation occurs, ft. (m)

ΔSG = difference in specific gravity between the oil and water relative to water

β_w = fraction of water height within the vessel = h/d

α_w = fractional cross-sectional area of water

Any combination of L_{eff} and d that satisfies the above settling criteria equation will be sufficient to allow all oil particles of diameter d_m or larger to settle out of the water.

The fractional water height and fractional water cross-sectional areas are related by the following

$$\alpha_w = \left(\frac{1}{180}\right) \cos^{-1}[1 - 2\beta_w] - \left(\frac{1}{\pi}\right)[1 - 2\beta_w]$$
$$\sin\left[\cos^{-1}(1 - 2\beta_w)\right] \qquad (1\text{-}7c)$$

After fractional water height within the vessel is selected, the associated fractional cross-sectional area may be calculated using 1-7c. These values may then be used in Equations 1-7a and 1-7b.

Retention time criteria:

In addition to settling criteria, a minimum retention time should be provided to allow coalescence.

As stated earlier, increasing retention time beyond that required for initial coalescence is not a cost-effective method for increasing oil droplet diameter.

However, some initial retention time can cost-effectively increase the oil droplet size distribution.

Retention times vary from 10 to 30 minutes.

It is recommended that a retention time of not less than 10 minutes be provided in vessels that have no means of promoting coalescence.

To ensure that the appropriate retention time has been provided, the following equation must also be satisfied when one selects d and L_{eff}.

Field units

$$d^2 L_{eff} = \frac{(t_w)_w Q_w}{1.4\alpha_w} \qquad (1\text{-}8a)$$

SI units

$$d^2 L_{eff} = 21,000 \frac{(t_r)_w Q_w}{\alpha_w} \qquad (1\text{-}8b)$$

where $(t_r)_w$ = retention time, min.

The choice of correct diameter and length can be obtained by selecting various values for d and L_{eff} for both Equation 1-8 and 1-9.

For each d, the larger L_{eff} must be used to satisfy both equations.

Seam-to-seam length:

The relationship between the L_{eff} and the seam-to-seam length of a skimmer depends on the physical design of the skimmer internals.

Some approximation of the seam-to-seam length may be made based on experience as follows:

$$L_{ss} = (4/3)L_{eff} \qquad (1\text{-}9)$$

where L_{ss} = seam-to-seam length, ft. (m)

This approximation must be limited in some cases, such as vessels with large diameters. Therefore, the L_{eff} should be calculated using Equation 1-9 but must be equal to or greater than the values calculated using the following equations:

Field units

$$L_{ss} = L_{eff} + 2.5 \qquad (1\text{-}10a)$$

SI units

$$L_{ss} = L_{eff} + 0.76 \qquad (1\text{-}10b)$$

Field units

$$L_{ss} = L_{eff} + (d/24) \qquad (1\text{-}11a)$$

SI units

$$L_{ss} = L_{eff} + (d/2000) \qquad (1\text{-}11b)$$

Equation 1-10 will govern only when the calculated L_{eff} is less than 7.5 ft. (2.3 m).

> The justification for this limit is that some minimum vessel length is always required for oil and water collection before dumping.

Equation 1-11 governs when one-half the diameter in feet exceed one-third of the calculated L_{eff}.

> This constraint ensures that even flow distribution can be achieved in short vessels with large diameters.

Horizontal Rectangular Cross-Section Skimmer

The required width and length of a horizontal tank of rectangular cross section can be determined from Stokes' law using an efficiency factor of 1.9 for turbulence and short-circuiting.

Settling criteria:

Field units

$$WL_{eff} = \frac{70Q_w\mu_w}{(\Delta SG)d_m^2} \qquad (1\text{-}12a)$$

SI units

$$WL_{eff} = \frac{950Q_w\mu_w}{(\Delta SG)d_m^2} \qquad (1\text{-}12b)$$

where

W = width, ft. (m)

L_{eff} = effective length in which separation occurs, ft.(m).

Equations 1-12a and 1-12b are independent of height.

This is because the oil settling time and the water settling retention time are both proportional to the height.

Typically, the height of the water flow is limited to less than one-half the width to assure good flow distribution.

With this assumption, the following equation can be derived to ensure that sufficient retention time is provided.

If the height-to-width ratio is set to 50%, then the following retention time equation applies.

Retention time criteria:

Field units

$$W^2 L_{eff} = \frac{0.004(t_r)_w Q_w}{\gamma} \qquad (1\text{-}13a)$$

SI units

$$W^2 L_{eff} = \frac{(t_r)_w Q_w}{60\gamma} \qquad (1\text{-}13b)$$

where:

γ = height-to-width ratio, H/W

The choice of W and L that satisfies both requirements can be obtained graphically.

The height of water flow, H, is set equal to 0.5 W.

Seam-to-seam length:

As with horizontal cylindrical skimmers, the relationship between L_{eff} and L_{ss} is dependent on the internal design.

Approximations of L_{ss} of rectangular skimmers may be made using Equation 1-9 and 1-10.

However, the L_{ss} must be limited by the following:

$$L_{ss} = L_{eff} + (W/20) \qquad (1\text{-}14)$$

As before, the L_{ss} should be the largest of Equations 1-9, 1-10, and 1-14.

Vertical Cylindrical Skimmer

One can determine the required diameter of a vertical cylindrical tank by setting the oil rising velocity equal to the average water velocity as follows:

Settling criteria

Field units

$$d^2 = \frac{6691 F Q_w \mu}{(\Delta SG) d_o^2} \qquad (1\text{-}15a)$$

SI units

$$d^2 = \frac{6.365 \times 10^8 F Q_w \mu_w}{(\Delta SG) d_o^2} \qquad (1\text{-}15b)$$

where

$F =$ factor that accounts for turbulence and short circuiting $= 1$ (diameters < 48 in. (1.2 m)) $= d/48$ (diameters > 48 in. (1.2 m))

$d =$ vessel diameter, inches

Substituting this into Equation 1-15 yields the following:

Field units

$$d^2 = \frac{140 F Q_w \mu_w}{(\Delta SG) d_o^2} \qquad (1\text{-}16a)$$

SI units

$$d^2 = \frac{5.3 \times 10^9 F Q_w \mu}{(\Delta SG) d_o^2} \qquad (1\text{-}16b)$$

Retention time criteria:

The height of the water column in a vertical skimmer can be determined for a selected d from retention time requirements:

Field units

$$H = \frac{(0.7)t_w Q_w}{d^2} \qquad (1\text{-}17a)$$

SI units

$$H = \frac{21,218 t_w Q_w}{d^2} \qquad (1\text{-}17b)$$

where

H = height of the water, ft. (m)

d = vessel diameter, in. (mm)

t_w = retention time, minutes

Q_w = water flow rate, bwpd (m³/hr)

Seam-to-seam length:

The height of the oil pad in both vertical and horizontal skimmers typically ranges from 2 to 6 in. (50 to 150 mm).

The purpose of the skimmer is to remove oil from the water and produce as clean a water stream as possible.

The quality of the skimmed oil from a skimmer is secondary consideration.

Skimmed oil streams typically contain 20 to 50% water.

The objective is to maximize the water treating ability of the skimmer.

Maintaining a minimum oil pad thickness accomplishes this objective.

Coalescers

General Considerations

Several different types of devices have been developed to promote the coalescence of small dispersed oil droplets.

These devices:

Use gravity separation similar to skimmers but also induce coalescence to improve the separation.

Either match the performance of a skimmer or offer improved performance in the same space.

Plate Coalescers

Various configurations of plate coalescers have been devised.

Common configurations:

Parallel plate interceptor (PPI)

Corrugated plate interceptor (CPI)

Cross-flow separators

All configurations depend on gravity separation to allow:

Oil droplets to rise to a plate surface

Where coalescence and capture occur

It is possible to overcome the size and weight disadvantage of skim tanks by enhancing coalescence of the oil droplets, thereby substantially increasing their rise velocities.

Plate coalescers require smaller cross-sectional areas, thus providing space and weight gains over skim tanks.

As shown in Figure 1-10:

Flow is split between a number of parallel plates spaced a ½ to 2 in. (1.2 to 5 cm) apart.

To facilitate capture of the oil droplet the plates are inclined to the horizontal which:

Promotes oil droplet coalescence into films

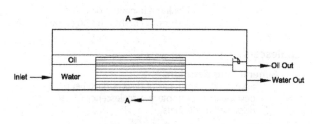

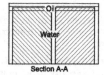

FIGURE 1-10 Schematic of a parallel plate interceptor.

Guides the oil to the top for entrapment into channels, thereby preventing remixing with the water.

Plates provide a surface for the oil droplets to collect and for solids particles to settle.

As shown Figure 1-11:

An oil droplet entering the space between the plates will

Rise in accordance with Stokes' law

Have a forward velocity equal to the bulk water velocity

By solving for the vertical velocity needed by a particle entering at the base of the flow to reach the coalescing plate at the top of the flow, the resulting droplet diameter can be determined.

Stokes' Law should apply to oil droplets as small in diameter as 1 to 10 microns.

**LARGE DROPLETS RISE
TO COLLECTION SURFACE**

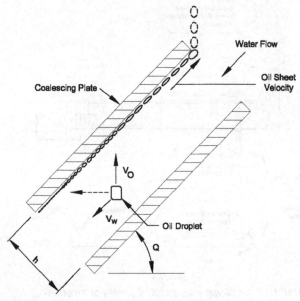

FIGURE 1-11 Cross section showing plate coalesce operation.

Field experience indicates that:

> 30 microns sets a reasonable lower limit on the droplet size that can be removed.

> Below 30 microns, small pressure fluctuations, platform vibration, and so on, tend to impede the rise of the droplets to the coalescing surface.

Parallel Plate Interceptor (PPI)

For the first form of a plate coalescer, refer to Figure 1-10.

PPIs involve installing a series of plates parallel to the longitudinal axis of an API separator (a horizontal rectangular cross section skimmer). Figure 1-12.

Plates form a "V" when viewed along the axis of flow so that the oil sheet migrates up the underside of the coalescing plate and to the sides.

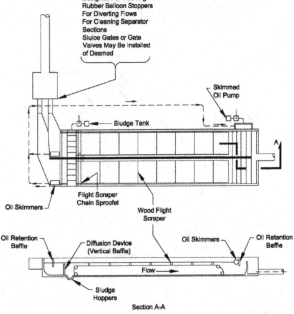

FIGURE 1-12 API oil/water separator. Courtesy of American Petroleum Institute.

Sediments migrate toward the middle and down to the bottom of the separator where they are removed.

Small inter-plate spacing:

> Allows packing more plates inside a vessel
>
> Maximizes the area for oil droplets to coalesce
>
> Increases the probability of plugging the inter-spaces with solids
>
> As a compromise, a distance of ¾ inch is typically used
>
> The angle of inclination for the plates is generally established at 45°.

Corrugated Plate Interceptor (CPI)

The corrugated plate interceptor is the most common form of parallel plate interceptor used in production operations.

It is a refinement of the PPI in that it:

> Takes up less plan area for the same particle size removal
>
> Makes sediment handling easier
>
> Is cheaper than a PPI

Figure 1-13 illustrates the flow pattern of a typical downflow CPI design.

> Water enters the inlet nozzle (1), where solids flow downward and settle in the primary collection box (2).
>
> Water and oil flow up and through a perforated distribution baffle plate (3).
>
> The CPI pack (4) receives oily water.
>
> Oil rises out of the flow path to the underside of the ridge and coalesces into a film moving upward opposite the bulk water flow.
>
> A thick layer of oil is allowed to collect until it flows over an adjustable weir (5) into an oil collection box for removal.
>
> Light solids and sludge separation is simultaneously accomplished and falls to the lower plate surface along the gutters and collectors at the bottom (6), where it is removed.

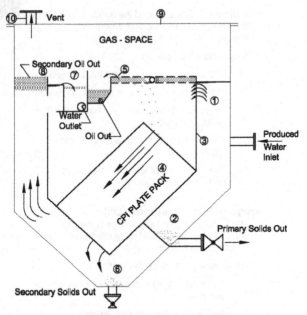

FIGURE 1-13 Schematic showing flow pattern of a typical downflow CPI design.

After exiting the CPI pack, the water moves upward and flows over an adjustable weir (7) into the water removable box.

A secondary oil removal outlet (8) is located above the water outlet.

A gasketed cover (9) allows for gas blanket operation.

It is also supplied with an adequately sized vent nozzle (10).

Parallel plates are corrugated, like roofing material, with the axis of the corrugations parallel to the direction of the flow.

Figure 1-14 shows a typical plate pack.

The plate pack, contained in a box-shaped frame, is tilted at a 45° angle and the bulk water flow is forced downward.

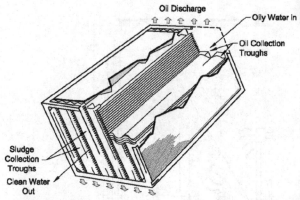

FIGURE 1-14 CPI plate pack.

> The oil sheet rises upward to the water flow
> and is concentrated in the top of each
> corrugation.

> When the oil reaches the end of the plate
> pack, it is collected in a channel and brought
> to the oil–water interface.

In areas where sand or sediment production is
anticipated, the sand should be removed prior to
flowing through a standard downflow CPI.

Because of the required laminar flow regime, all
plate coalesces are efficient sand settling devices.

Field experience has shown:

> Oil-wet sand may adhere to the standard 45°
> sloped plate and clog.

> Sand collection channels installed at the end
> of the plate pack cause turbulence that affects
> the treating process and are themselves
> subject to sand plugging.

To eliminate the above problems, an "upflow" CPI
unit may be used (Figure 1-15).

> Uses corrugated plates, spaced a minimum of
> 1 in. (2.5 cm) apart

> With a 60° angle of inclination.

Figure 1-16 compares the flow pattern of an
upflow and downflow CPI pack.

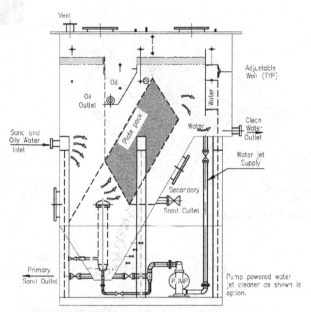

FIGURE 1-15 Schematic showing flow pattern of a typical upflow CPI design.

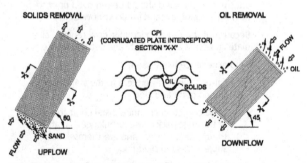

FIGURE 1-16 Upflow versus downflow pattern.

The main components of a CPI plate separator are:

Separator basin

CPI plate pack

Oil and effluent weir

Basin cover

Solids hopper

Inlet and outlet nozzles

Separator basin and its internals have the following characteristics:

Made of carbon steel plate with at least 3/16-in thickness

Basin edges are welded

All carbon steel external and internal surfaces are blast-cleaned and painted with epoxy paint.

CPI plate packs:

Materials of construction include:

Chlorinated polyvinylchloride (CPVC)

Polyvinylchloride (PVC)

Polypropylene (PP)

Fiberglass reinforced polyester

Carbon steel

Galvanized steel

Various grades of stainless steel

Temperature limitations:

Polymer plates are limited to 140°F (55°C)

Stainless steel are limited to 350°F (125°C)

Plate pack usually has a 316 SS frame for robustness and easy removal during maintenance.

Polypropylene plates:

Have an inherently oleophilic property that attracts oil, thus promoting coalescence

Repels water, which adds the downward flow of sludge, thus reducing chances of sludge fouling

Oil weir is a bucket type and made of carbon or stainless steel

Effluent weir is a plate type and its height is adjustable.

Basin cover is normally made of carbon steel, heavy-duty galvanized steel, or lightweight fiberglass reinforced plastic (FRP) with ³/₁₆-in. thickness.

Solids hopper may be conical or dish-shaped for cylindrical separators, or shaped like an inverted pyramid for rectangular separators.

Vessel should be leak-tested prior to coating. The assembled package should be dry function tested to ensure proper operation. Any plastic piping should also be hydrotested.

Cross-Flow Devices

A cross-flow device is a modified CPI configuration where water flows perpendicular to the axis of the corrugations in the plates (see Figure 1-17).

It allows plates to be put on a steeper angle to facilitate sediment removal

It enables plate pack to be packaged into a pressure vessel, thus providing protection against gas blow-by

It can be configured in either horizontal or vertical pressure vessels.

Horizontal Units (Figure 1-18)

Horizontal units require less internal baffling as the ends of each plate conduct the

Oil to the oil-water interface

Sediments to the sediment area below the water flow area

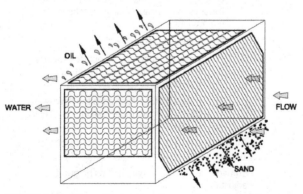

FIGURE 1-17 Schematic showing flow pattern of cross-flow plate pack.

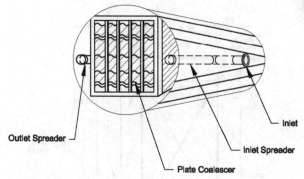

Outlet Spreader

Inlet

Inlet Spreader

Plate Coalescer

FIGURE 1-18 Schematic showing cross-flow device installed in a horizontal pressure vessel.

Pack is long and narrow, and thus requires an elaborate spreader and collection device to force the water to travel across the plate pack in plug flow.

Inlet oil droplets may shear in the spreader, which would make separation more difficult.

This configuration would be preferred when a pressure vessel in a high-pressure system is needed.

Vertical Units

Vertical units require collection channels on both ends:

One end to enable the oil to rise to the oil/water interface

One end to allow the sand to settle to the bottom

Can be designed for more efficient sand removal.

Cross-flow device may be installed in an atmospheric vessel, as shown in Figure 1-19, or in a vertical pressure vessel.

CPI separators are generally cheaper and more efficient at oil removal than cross-flow separators.

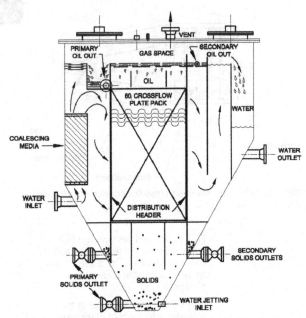

FIGURE 1-19 Schematic showing cross-flow device installed in an atmospheric vessel.

Cross-flow separators should be considered where:

A pressure vessel is preferred

High sand production is expected, and

Sand is not removed upstream of the water treating equipment

Performance Considerations

Flow direction considerations include:

Downflow

Efficient for oil removal

Plate pack inclined at a 45° angle

Upflow

Used when the flowstream contains significant amounts of solids

Plate pack inclined at a 60° angle

Higher plate slope provides about 25% greater runoff force and a 30% lower erosion rate than a 45° plate slope

Cross Flow

Should be considered where the use of a pressure vessel is preferred and solids and oil removal is desired.

Plate separators exhibit the following advantages:

Require very little maintenance

Packs can be easily removed as complete modules for inspection and cleaning.

Smaller in size and weight than skim vessels because of the effect of the closely spaced inclined plates

Can accept inlet feed oil concentrations as high as 3000 mg/l

Can separate oil droplets down to about 30 microns

Have a sand removal ratio of 10:1

If unit captures 50-micron oil droplets, it will also capture solid particles as low as 5-microns.

Totally enclosed, thereby eliminating vapor losses and reducing fire hazards

CPIs are more efficient at oil removal than cross-flow separators

Simple and expensive in comparison to other types of produced water treating devices, that is, flotation units

Have no moving parts and do not require power

Easy to cover due to their size and retain hydrocarbon vapors

Easy to install in a pressure vessel

Helps to retain hydrocarbon vapors

Protect against overpressure due to failure of an upstream level control valve

Disadvantages include:

Not effective for streams with slugs of oil

Cannot effectively handle large amounts of solids and emulsified streams

Plate separators are recommended when:

Water flow rate is steady or feed is from a pump

Size and weight are not constraints

Utilities and equipment are available to periodically clean the plate packs

Influent oil content is high and oil concentrations must be reduced to 150 mg/l for effective second-stage treating in a downstream unit

Solid contaminants are not significant in the waste stream and sand content is less than 110 ppm

Plate separators are not recommended when:

Influent droplet sizes are mostly below 30-microns

Size and weight are the primary considerations

Sand particle diameters are less than 25 microns and solids removal is a primary objective

Selection Criteria

Plate separators are effective to approximately 30 microns.

Vendor-supplied nomographs can be used to estimate the performance of CPIs.

Figure 1-20 is a nomograph for a "downflow" CPI unit.

Shows relationship among:

Liquid inflow temperature

Particle size removal

Differential specific gravity of the oil and water, and

Capacity for downflow oil removal

An example using Figure 1-20:

Produced water flowing at a rate of 150 gpm (5143 bbl/day) per CPI pack with ¾-inch

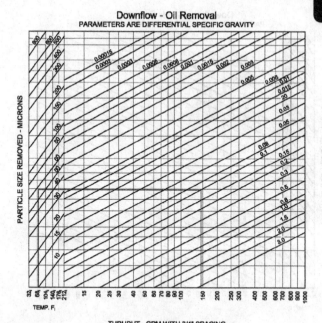

Downflow - Oil Removal
PARAMETERS ARE DIFFERENTIAL SPECIFIC GRAVITY

PARTICLE SIZE REMOVED - MICRONS

TEMP. F

THRUPUT - GPM WITH 3/4" SPACING
CPI or VPI standard size
45, PLATE ANGLE

FIGURE 1-20 Nomograph for downflow CPI.

spacing, a differential specific gravity of 0.1, and a flowing temperature of 68° F will remove a particle of about 60 microns.

Figure 1-21 is a nomograph for "upflow" solids and oil removal.

Figure 1-22 is a nomograph for a "cross flow" CPI unit.

Coalescer Sizing Equations

The general sizing equation for a plate coalescing with flow either parallel to or perpendicular to the slope of the plates for droplet removal is:

Field units

$$HWL = \frac{4.8 Q_w h \mu_w}{COS\theta (d_o)^2 (\Delta SG)} \qquad (1\text{-}18a)$$

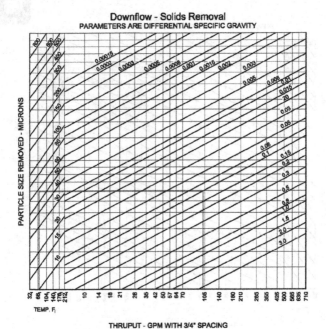

Downflow - Solids Removal
PARAMETERS ARE DIFFERENTIAL SPECIFIC GRAVITY

THRUPUT - GPM WITH 3/4" SPACING
CPI or VPI standard size
60, PLATE ANGLE

FIGURE 1-21 Nomograph for upflow CPI.

SI units

$$HWL = \frac{0.794 Q_w h \mu_w}{COS\theta(d_o)^2(\Delta SG)} \qquad (1\text{-}18b)$$

where

d_o = design oil droplet diameter, microns

Q_w = bulk water flow rate, bwpd (m^3/hr)

h = perpendicular distance between plates, in. (mm)

μ = viscosity of water

θ = angle of the plate with the horizontal

H, W = height and width of the plate section perpendicular to the axis of water flow, ft. (m)

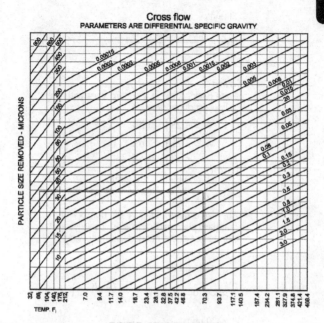

Cross flow
PARAMETERS ARE DIFFERENTIAL SPECIFIC GRAVITY

THRUPUT - GPM WITH 3/4" SPACING
CPI or VPI standard size
60₁ PLATE ANGLE

FIGURE 1-22 Nomograph for cross-flow CPI.

L = length of plate section parallel to the axis of water flow, ft. (m)

ΔSG = difference in specific gravity between the oil and water (relative to water = 1)

Experiments have indicated that Reynolds number for the flow regime cannot exceed 1600 with four times the hydraulic radius as the characteristic dimension.

Based on this correlation, the minimum H times W for a given Q_w can be determined from:

Field units

$$HW = \frac{14 \times 10^4 Q_w h(SG)_w}{\mu_w} \qquad (1\text{-}19a)$$

SI units

$$HW = \frac{8.0 \times 10^4 Q_w h (SG)_w}{\mu_w} \qquad (1\text{-}19b)$$

CPI Sizing

Plate packs come in standard sizes with the following dimensions:

$H = 3.25$ ft. (1 m)

$W = 3.25$ ft. (1 m)

$L = 5.75$ ft. (1.75 m)

$h = 0.69$ in. (190 mm)

$\theta = 45°$

The size is determined by the number of standard plate packs installed. To arrive at the number of packs needed, the following equation is used:

Field units

$$\text{Number of packs } (N) = \frac{0.77 Q_w \mu}{(\Delta SG) d_0^2} \qquad (1\text{-}20a)$$

SI units

$$\text{Number of packs } (N) = \frac{11.67 Q_w \mu}{(\Delta SG) d_0^2} \qquad (1\text{-}20b)$$

Where:

$N =$ number of plate packs

To ensure the Reynolds number limitation of 1600 is met to prevent turbulence, the flow through each pack should be limited to approximately 20,000 bwpd (130 m³/hr).

Similarly, the flow through a half pack unit should be less than 10,000 bwpd (66 m³/hr).

These flow rate limits are for the maximum flow, and any surging should be considered.

To allow for surging it is recommended that the average flow through each full pack be less than 10,000 bwpd (66 m³/hr).

Higher average flows may be used, but care must be used in designing the system to avoid surges that might exceed the 20,000 bwpd (130 m³/hr) limit for short periods of time.

In areas where sand or sediment production is anticipated, the sand should be removed prior to flowing through a standard CPI.

Because of the required laminar flow regime, all plate coalescers are efficient sand settling devices.

Experience has shown that oil-wet sand may adhere to a 45° slope, which creates the possibility that the sand will adhere to and clog the plates.

In addition, the sand at the end of the plate pack causes turbulence, which affects the treating process; plate coalescers are also susceptible to sand plugging.

To help alleviate the solids plugging problem inherent in CPIs, it is possible to specify a 60° angle of inclination which increases the number of packs to 40% according to the following:

Field units

$$\text{Number of packs} = \frac{0.11 Q_w \mu}{(\Delta SG) d_0^2} \qquad (1\text{-}21a)$$

SI units

$$\text{Number of packs} = \frac{16.68 Q_w \mu}{(\Delta SG) d_0^2} \qquad (1\text{-}21b)$$

Cross-Flow Device Sizing

Cross-flow device sizing obeys the same general sizing equations as plate coalescers.

Some manufacturers claim greater efficiency than CPIs, the reason for this is not apparent from theory, laboratory/field tests, or personal experience.

If the height and width of these cross-flow packs are known, Equation 1-19a or 1.19b can be used directly.

It may be necessary to include an efficiency term, normally 0.75, in the denominator on the right side of

Equation 1-19 if the dimensions of H or W are large and a spreader is needed.

Both horizontal and vertical cross-flow separators require spreaders and collectors to uniformly distribute the water flow among the plates.

For this reason, the following equation has been developed assuming a 75% spreader efficiency term:

Field units

$$HWL = \frac{(6.4)Q_w h \mu_w}{\cos\theta (\Delta SG) d_0^2}$$ (1-23a)

SI units

$$HWL = \frac{(1.06)Q_w h \mu_w}{\cos\theta (\Delta SG) d_0^2}$$ (1-23b)

Example 1-1: Determining the Dispersed Oil Content in the Effluent Water from a CPI Plate Separator

Given:

Feed water flow rate   = 25,000 bwpd @ 125°F

Feed water specific gravity = 1.06 @ 125°F

Feed water viscosity   = 0.65 cp @ 125°F

Dispersed oil concentration = 650 mg/l

Dissolved oil concentration = 10 mg/l

Total oil and grease   = 660 mg/l

The dispersed oil droplet size distribution in feed water is as follows:

Microns:	<40	40–60	60–80	80–100	100–120	>120
Vol % :	914	30	35	10	2	0

A vendor has quoted that one of its standard plate packs would be capable of reducing the total oil and grease content of the effluent water to less than 200 mg/l.

The vendor's standard plate pack has the following geometric specifications:

$H = 3.25$ ft.

$W = 3.25$ ft.

$L = 5.75$ ft.

$h = 0.69$ in.

$\theta = 45°F$

Determine: Calculate the total oil and grease content in effluent water from the plate pack to check the vendor's quoted proposal.

Solution: In order to calculate the total oil and grease in the effluent water, one must determine the smallest oil droplet size that can be removed in the vendor's standard plate pack at the design conditions given.

Equation 1-21 was derived for the given plate pack geometric configuration assumed in the example calculation.

Rearranging Equation 1-21 to solve for the minimum oil droplet size (d_o), we have the following results:

$$d_o = \sqrt{\frac{0.77 Q_w \mu}{\Delta SG (of packs @ 45°)}}$$

$$= \sqrt{\frac{0.77(25,000)(0.65)}{(1.06 - 0.75)(1)}}$$

$$= 63.5 \text{ microns}$$

The volume percent of the dispersed oil by the plate pack is determined by summing the volume percent of dispersed oil droplets contained in the feed water that are greater than or equal to 63.5 microns (see dispersed oil droplets size distribution data given). Therefore,

$$\text{Vol. \% removed} = \frac{(80 - 63.5)}{(80 - 60)}(30) + (35)+$$

$$(10) + (2)) = 71.75\%$$

Calculating the dispersed oil content in the effluent water from the plate pack (C_{out}):

$C_{out} = (650)(100\% 71.75\%)$
$= 183.6 \text{ mg/l}$

Since the plate pack does not remove any of the dissolved oil, the total oil and grease content in the effluent water from the plate pack is equal to 183.6 mg/l plus 10 mg/l, or 193.6 mg/l.

Thus, the vendor's quoted performance looks to be correct.

Oil/Water/Sediment Coalescing Separators

Enhancement of the cross-flow configuration in that it utilizes a two-step process to separate small oil droplets and solids from the well stream.

The coalescing packs used are cross flow in design rather than downflow or upflow.

The units can be configured in either an atmospheric pressure tank (see Figure 1-23) or a vertical pressure vessel (see Figure 1-24).

Both configurations use:

Inlet flow distributer/coalesce pack

Cross-flow plate pack

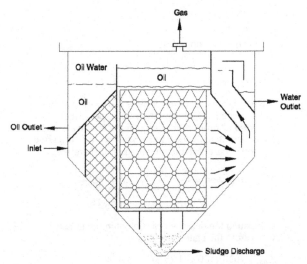

FIGURE 1-23 Schematic of an oil/water/settlement coalescing tank.

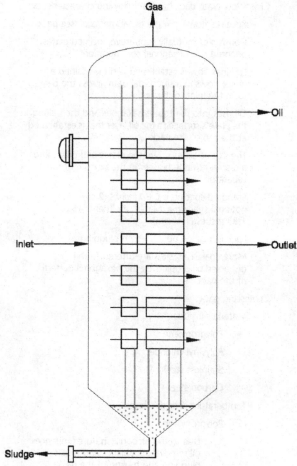

FIGURE 1-24 Schematic of an oil/water/settlement coalescing vessel.

Inlet flow distributer/coalesce pack has the following characteristics:

Evenly spreads the inlet flow over the full height and width of the separator pack

Flow through the pack is mildly turbulent thus creating opportunities for the oil droplets to coalesce into larger ones

Cross-flow plate pack has the following characteristics:

Receives flow from the distributer/coalesce pack

Consists of mutually supportive, inclined plates oriented in a hexagonal configuration

Laminar flow is established and maintained as water flows in a sinusoidal path across the pack from the inlet to the outlet

Oil rises into the top of hexagons and then along the plate's surface to the oil layer that is established at the top of the pack

The sludge slides down the plates and drops into a discreet sludge hopper in the bottom of the separator

Standard spacing is 0.80 inches (2 cm) with optional spacing available of either 0.46 or 1.33 inches (1.17 cm)

Pack is inclined 60° to lessen plugging

More coalescing sites are offered to the dispersed oil droplets by the hexagonal pattern of the pack

Coalescing pack:

Materials include:

Polypropylene

Polyvinyl chloride

Stainless steel

Carbon steel

Temperature considerations:

Polypropylene

Due to the oleophilic nature (enhances oil removal capabilities and resists plugging and fouling of the pack)

Used up to 150°F (66°C) as pack loses integrity and chemical degradation begins above this temperature

Stainless steels and carbon steels

Used in temperatures above 150°F (66°C)

Used in environments that contain large amounts of aromatic hydrocarbons

Oil/Water/Sediment Coalescing Separator Sizing

The geometry of plate spacing and length can be analyzed for this configuration using Equation 1-19 and techniques previously discussed.

Performance Considerations

Exhibits the same advantages and disadvantages as plate separators.

One additional improvement is that the minimum oil droplet size that can be removed is 20 microns.

Skimmers/Coalescers

Several designs that are marketed for improving oil-water separation:

Rely on installing coalescing plates or packs within horizontal skimmers or free-water knockouts

Encourage coalescence and capture of small oil droplets within the water continuous phase

Coalescers:

Accumulate oil on a preferentially oil wet surface where small droplets can accumulate late

These larger oil droplets can be either

Collected directly from the oil wet surface, or

Stripped from the oil wet surface and separated from the water phase using some type of gravity-based equipment

Coalescing equipment may either be:

Housed in a separate vessel, or

More commonly, installed in a coalescing pack contained in a gravity vessel

Figure 1-25 shows a schematic of a horizontal FWKO with a coalescing pack.

The plates of a CPI or cross-flow vessel may be fabricated of an oleophilic (oil-wetting) material and thereby serve as both as a:

Gravity separation device, and

Coalescing device

Figure 1-26 is a cross section of structured packing serving as a coalescer.

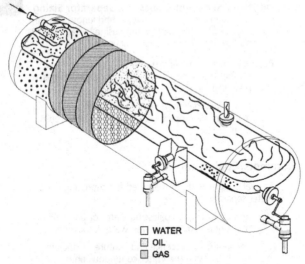

☐ WATER
☐ OIL
☐ GAS

FIGURE 1-25 Schematic of an FWKO with a coalescing pack.

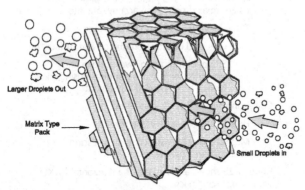

FIGURE 1-26 Structured packing serving as a coalesce.

Matrix Type

Mats or fibers have the advantage of large surface areas and easy fabrication.

Oleophilic materials are spun into thin fibers, and the fibers are collected into a pack, across which the oily water flows.

FIGURE 1-27 Oil coalescence on a fibrous mat.

Oil droplets stick to the fibrous mat.

Figure 1-27 illustrates the coalescence process on a fibrous mat.

Coalesced droplets can easily be collected after they emerge from the mat using a gravity-based separator (see Figure 1-28).

Loose Media

Oleophilic material can also be fabricated into loose media, and the media collected in a vessel.

If the material is fabricated in a granular form and assembled into a deep bed gravity settle, the deep-bed filter can also perform a coalescing function.

The geometry of plate spacing and length can be analyzed for each of these designs using Equation 1-19 and the techniques previously discussed.

The packs cover the entire inside diameter of the vessel unless sand removal internals are required.

Pack lengths range from 2 to 9 ft. (0.6 to 3.8 m), depending on service.

Performance Considerations

Coalescers:

Are used to improve the performance of other gravity-based separation equipment

Are either:

Specified by the equipment vendor as an integral part of the water treating system, or

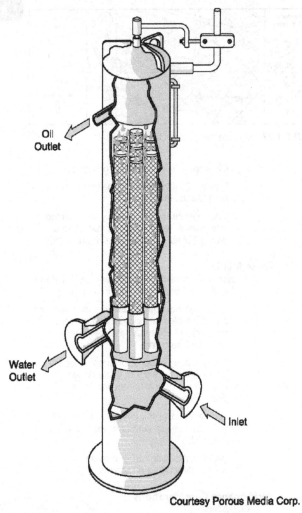

Courtesy Porous Media Corp.

FIGURE 1-28 Collection of oil from a matrix separator.

Added as a retrofit to improve the performance of an existing system

Are particularly useful when the oil droplet size in the incoming water is small as a result of excess shearing in upstream piping or valves.

Coalescers can be used when:

> Existing low-pressure separator, skimmer, or plate separator can be retrofitted with a coalescing section

> Coalescing section is accessible for cleaning and replacement

> Inlet oil droplet size is less than 50 microns and larger droplets are desired

Coalescers can also serve as a skimmer (the limitations listed for skimmers are applicable).

Coalescers should not be used when:

> Inlet droplet sizes are less than 10 microns

> Inlet droplet sizes are greater than 100 microns

> Size and weight are primary considerations

Precipitators/Coalescing Filters

Precipitators are obsolete and would not be used in a new installation.

In the past, it was common to direct the water to be treated through a bed of excelsior (straw) or another similar medium, as shown in Figure 1-29, to aid in the coalescing of oil droplets.

Coalescing medium has a tendency to clog.

Many of these devices in oilfield service have the medium removed.

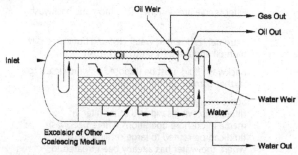

FIGURE 1-29 Schematic of a precipitator.

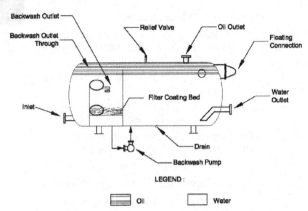

FIGURE 1-30 Schematic of a coalesce.

They act like a vertical skimmer since the oil droplets must flow countercurrent to the downward flow of the water through the area where the medium was originally located.

Coalescers as shown in Figure 1-30 have the following characteristics:

They are similar in design to a precipitator except that they:

Usually employ a larger gravity separation section than a precipitator

Utilize a back-washable filter bed for coalescing and some sediment removal

Filter media are designed for automatic backwash cycles

Extremely efficient at water cleaning but clog easily with oil and are difficult to backwash

Backwash fluid must be disposed of which leads to further complications

Some operators have had success with filters employing sand and other filter media in onshore operations where backwash fluid can be routed to large settling tanks, and where the water has already been treated to 25–75 mg/l oil

Applications of this type are typical when the
produced water will be reinjected as for
water-flood

Free-Flow Turbulent Coalescers

Plate coalescing devices:

Use gravity separation followed by coalescence to
treat water

Have the disadvantage of requiring laminar flow
and closely spaced plates in order to capture the
small oil droplets and keep them from stripping the
coalesced sheet

Are susceptible to plugging with solids

Free-flow turbulent coalescers:

Are devices installed inside or just upstream of any
skim tank or coalesce to promote coalescence

Were marketed and sold under the trade name
SP Packs

They are no longer available for sale but the
concept can still be employed in water
treating design.

SP Packs (Figure 1-31):

Force the water flow to follow a serpentine
pipe-like path sized to create turbulence of
sufficient magnitude to promote coalescence,
but not so great as to shear the oil droplets
below a specified size

Are less susceptible to plugging since they:

Require turbulent flow (high Reynolds
numbers)

Have no closely spaced passages

Have a pipe path similar in size to the inlet
piping

Are designed to coalesce oil droplets to a
defined drop size distribution, with a d_{max} of
1000 microns

Are created by sizing a series of short runs
of pipe:

Diameters sized to create a Reynolds number
of 50,000 containing 6–10 short radius
180° bends

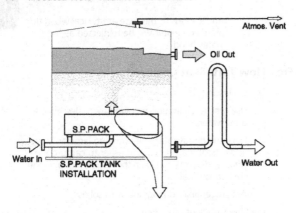

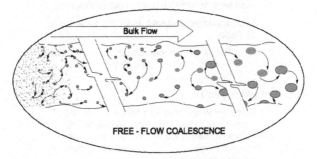

FIGURE 1-31 Principles of operation of an SP Pack.

Each run of straight pipe should be 30 to 50 pipe diameters

Increasing the d_{max} from a typical value of 250 microns in a normal inlet to a skim tank or coalesce to 1000 microns significantly reduces the size of the skimmer required

Effects on skimmer retention time:

May be reduced to 3 to 10 minutes as the need for retention time is not as important as the coalescence has occurred prior to the skimmer

SP Packs may be multistaged (Figure 1-32).

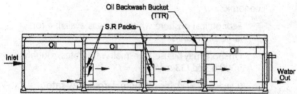

FIGURE 1-32 SP Packs installed in a horizontal flume.

As shown in Figure 1-33, a two-stage system may consist of:

SP Pack

Skim vessel, and

Second SP Pack

One SP Pack and skimmer combination constitutes one stage of coalescence and separation.

The second SP Pack coalesces the small oil droplets in the first skimmer's outlet, then the second skimmer may remove the larger oil droplets.

The addition of the SP Pack greatly improves the oil removal in the second skimmer because of coalescence.

If the second SP Pack were not in the system, all the large oil droplets would be removed in the first skimmer and the second skimmer would remove little oil.

Any number of stages in series may be used in the system.

SP Packs may be used as retrofit components to improve the performance of existing water treating systems.

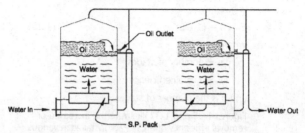

FIGURE 1-33 SP Packs installed in a series of staged tanks.

Economics:

> Economical onshore where space is available for large skim tanks
>
> Offshore may be used for small water rates, roughly 5000 bwpd (33 m³/hr)
>
> If space is available offshore, larger flow-rate applications may prove economical

As shown in Figure 1-34, the SP Pack may be placed inside any gravity settling device (skimmers, clarifiers, plate coalescers, etc.) and by growing a larger drop size distribution, the gravity settler is more efficient at removing oil, as shown in Figure 1-35.

Performance Considerations

The efficiency of each stage is given by:

$$E = \frac{C_i - C_0}{C_i} \tag{1-33}$$

where

C_i = inlet concentration

C_0 = outlet concentration

Since the drop size distribution developed by the SP Pack can be conservatively estimated as a straight line,

$$E = 1 - \frac{d_m}{d_{max}} \tag{1-24}$$

where

d_m = drop size that can be treated in the stage

d_{max} = maximum size drop created by the SP Pack = 1000 microns (for standard SP Packs)

The overall efficiency of a series staged installation is given by:

$$E_t = 1 - (1 - E)^n \tag{1-25}$$

where

n = number of stages

E_t = overall efficiency

and n is the number of stages

Figures 1-36 and 1-37 illustrate the increased oil removal efficiency of an SP Pack installed in various sized tanks.

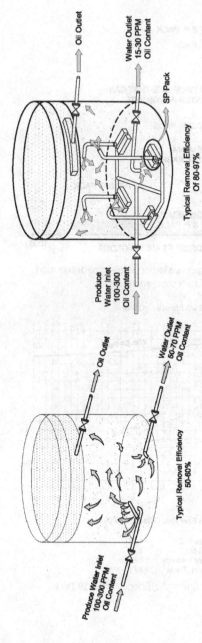

FIGURE 1-34 SP Packs installed in a clarifier skim tank. *Top*: before; *bottom*: after.

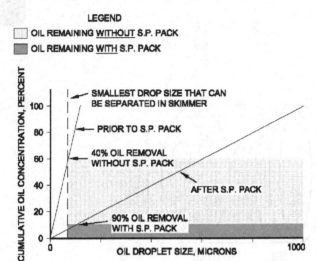

FIGURE 1-35 The SP Pack grows a larger droplet size distribution, thus allowing the skimmer to recover more oil.

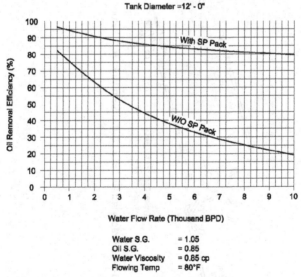

FIGURE 1-36 Improved oil removal efficiency of an SP Pack installed in a 12'-0" tank.

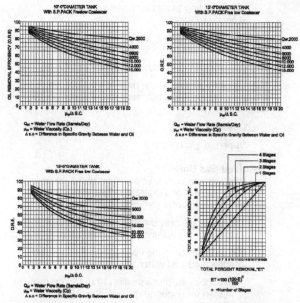

FIGURE 1-37 Oil removal efficiencies of various size tanks.

Flotation Units

Flotation units are the only commonly used water treating equipment that do not rely on gravity separation of the oil droplets from the water phase.

They employ a process in which fine gas bubbles have the following characteristics:

Generated and displaced in water

Attach themselves to oil and/or solid particulates

Rise to the vapor-liquid interface as oily foam, which is then skimmed from the water interface, recovered and then recycled for further processing

Effective specific gravity of the oil–gas bubble combination is significantly lower than that of a standalone oil droplet.

According to Stokes' law:

Resulting rising velocity of the oil–gas bubble combination is greater than that of a standalone oil droplet acting to accelerate the oil–water separation process.

Flotation aids such as coagulants, polyelectrolytes, or demulsifiers are added to improve performance.

Two distinct types of flotation units have been used.

They are distinguished by the method employed in producing the small gas bubbles needed to contact the water, specifically:

Dissolved gas units

Dispersed gas units

Dissolved Gas Units

Dissolved gas units take a portion of the treated water effluent and saturate the water with natural gas in a high-pressure "contactor" vessel.

The higher the pressure, the more gas can be dissolved in the water.

Gas bubbles are formed by flashing dissolved gas into the produced water.

Bubbles are much smaller (10 to 100 microns) than for dispersed gas flotation (100 to 1000 microns).

Gas volumes are limited by the solubility of the gas in water and are much lower than for dispersed gas flotation.

Units are designed for 20 to 40 psig (140 to 280 kPa) contact pressure.

Normally, 20 to 50% of the treated water is recirculated for contact with the gas.

Gas saturated water is then injected into the flotation tank as shown in Figure 1-38.

The dissolved gas breaks out of the oily water solution when the water pressure is flashed (reduced) to the low operating pressure of the flotation unit, in small-diameter bubbles that contact the oil droplets in the water and bring them to the surface in froth.

This type of flotation unit typically has not worked well in the oilfield.

They have been used successfully in refinery operations where:

Air can be used as the gas

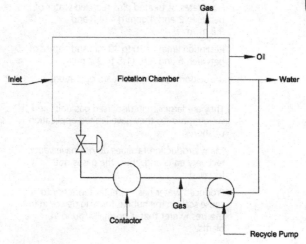

FIGURE 1-38 Schematic of a dissolved gas flotation process system.

Large areas are available for the equipment

Water to be treated is, for the most part, oxygenated fresh water

In treating produced water, it is desirable to use natural gas to:

Exclude oxygen

Avoid creating an explosive mixture

Minimize corrosion and bacteria growth

Using natural gas requires:

Venting of the gas, or

Installation of a vapor recovery unit

High dissolved solids content of produced water creates scale problems in these units.

Field experience with dissolved gas units in production operations have not been as successful as experienced with dispersed gas units.

Design parameters typically specified:

0.2 to 0.5 scf/bbl (0.036 to 0.89 Sm^3/m^3) of water to be treated

Flow rates of treated plus recycled water of between 2 and 4 gpm/ft.^{2} (4.8 and 9.8 m^3/m^3)

Retention times of 10 to 40 min and depths of between 6 and 9 ft. (1.8 to 2.7 m)

They are seldom used in upstream operations because:

They are larger than dispersed gas units and weigh more, so they have limited application offshore.

Many production facilities do not have vapor recovery units and, thus, the gas is not recycled.

Produced water has a greater tendency to cause scale in the bubble-forming device than the freshwater that is normally found in plants.

Dispersed Gas Units

Gas bubbles are dispersed in the total stream by the use of a:

Hydraulic inductor device, or

Vortex set up by mechanical rotors

Many different proprietary designs.

All require a:

Means to generate gas bubbles of favorable size and distribution into the flow stream

Two-phase mixing region that causes a collection to occur between the gas bubbles and the oil droplets

Flotation or separation region that allows the gas bubbles to rise to the surface

Means to skim the oily froth from the surface

Figure 1-39 shows the regions in which the above four processes occur.

These processes and the regions in which they occur are as follows:

Gas circulation path (A) and fluid circulation path (B) (bubble generation)

Two-phase mixing region (1) (attachment of oil droplets to the bubbles)

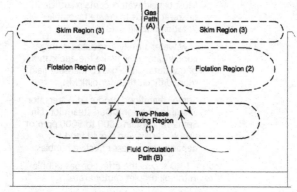

FIGURE 1-39 Dispersed gas flotation cell mechanics.

Flotation (separation) region (2) (rise of the bubbles to the surface applying Stokes' law)

Skimming region (3) (bubble collapse and oil skimming)

Gas bubble/oil droplet attachment can be enhanced with the use of polyelectrolyte chemicals.

Flotation aid chemicals can also be used to cause bubble/solid attachments, and thus flotation units can be used to remove solids as well.

Chemicals are added to the water to yield a chemical concentration between 1 and 10 ppm.

If oil has emulsifying tendencies de-emulsifiers may also have to be added in concentrations in the 20 to 50 ppm range.

Chemical treatment programs are highly location specific, and an effective treatment for one oil–water system may not be effective for another.

Units must generate a large number of small gas bubbles to operate efficiently. Tests indicate:

Bubble size decreases with increasing salinity

At salinities above 3%, bubble size appears to remain constant, but oil recovery often continues to improve.

Most oilfield waters contain sufficient dissolved solids to create favorable flotation bubble sizes.

Low water salinity associated with gas condensates may make the application of gas flotation to gas condensate fields more difficult than for oilfields.

Some steamflood produced waters, for example Chevron's Duri steamflood in Sumatra, contain 2000 to 5000 ppm of dissolved salts and would tend to generate large, less effective bubbles.

Figure 1-40 shows the effect of gas bubble size on the oil droplet capture rate.

Typical mean bubble sizes range between 50 to 60 microns.

Oil removal is dependent to some extent on oil droplet size.

Flotation has very little effect on oil droplets that are smaller than 2 to 5 microns.

It is important to avoid subjecting the influent to large shear forces

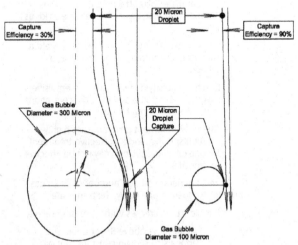

FIGURE 1-40 Effect of gas bubble size on oil droplet rate.

(e.g., throttling globe level control valves) immediately upstream of the unit.

It is best to separate control devices from the unit by long lengths of piping (at least 300 pipe diameters) to allow pipe coalescence to increase droplet diameter before flotation is attempted.

Above 10 to 20 microns, the size of the oil droplet does not appear to affect oil recovery efficiency, and thus elaborate inlet coalescing devices are not needed.

Field tests demonstrate that oil removal improves as the cumulative gas-water ratio increases. Table 1-3 shows the effects of installing multiple cells in series.

The rise and separation of oily bubbles from water require a relatively quiescent zone so that bubbles are not mixed into the bulk fluid.

Rising velocity of the bubbles must exceed both turbulent velocities and any downward bulk velocity.

Oily bubbles rise to the surface as an oily foam, which is then skimmed from the surface of the bulk water phase.

The skimming process acts to collapse the foam, which further concentrates the oily phase.

Skimming is usually achieved by a combination of weirs and skin paddles that move the oily foam to the edge of the cell and over the weir.

Table 1-3 Effects of Increased Gas Concentration on Oil Recovery

Water Location in Machine	Cumulative Gas/ Water Ratio, ft.^{3}/bbl	PPM Oil in Treated Water
Inlet water	0	225
Cell no. 1 effluent	8.8	97
Cell no. 2 effluent	17.5	50
Cell no. 3 effluent	26.3	18
Cell no. 4 effluent	35.0	15
Discharge cell effluent	35.0	14

The weir height relative to the position and speed of the skim paddles must be adjusted to prevent both excessive foam buildup on the bulk water surface and excessive water carryover into the oil skim bucket located below the weir.

Hydraulic Induced Units

Hydraulic induced units induce gas bubbles by gas aspiration into the low-pressure zone of a venture tube.

Figure 1-41 shows the flow path through a schematic cross section of a hydraulic induced unit.

Clean water from the effluent is pumped to a recirculation header (E) that feeds a series of Venturi educators (B).

Water flowing through the educators sucks gas from the vapor space (A) that is released at the nozzle (G) as a jet of small bubbles.

The bubbles rise, causing flotation in the chamber (C), forming a froth (D) that is skimmed with a mechanical device at (F).

Units are available in one, three, or four cells.

Figure 1-42 shows the flow path through a three-cell unit.

Four-cell units are designed with a 5-minute residence time based on normal capacity.

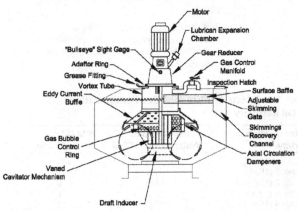

FIGURE 1-41 Schematic of a hydraulic induced gas flotation unit.

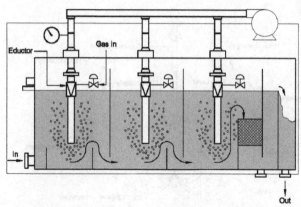

FIGURE 1-42 Schematic showing the flow path through a hydraulic induced flotation unit.

They use less power and less gas than mechanical rotor unit units.

They are less complex than the mechanical induced units.

Gas-water ratios less than 10 ft.^{2}/bbl at the design flow rate are used.

> Volume of gas dispersed in the water is not adjustable, so throughputs less than design result in higher gas-water ratios

Required water recycle rate to drive the educator varies with both the design capacity of the unit and between different manufacturers, but is generally around 50%

Since the water recirculated to the educators is 150% of the rated capacity, the average water throughput of the unit is three times.

Figure 1-43 is a sketch of an educator.

> Educator design is proprietary, and

> Varies considerably in both hydraulic design and mechanical placement between manufacturers

Control on bubble size and distribution is much more difficult than for mechanical units.

Stage efficiencies have a tendency to be lower than those of mechanical units.

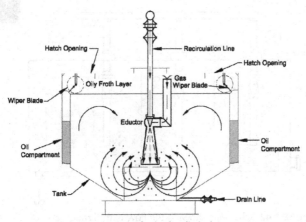

FIGURE 1-43 Cross section of a hydraulic inductor.

Mechanical Induced Units

Induce gas bubbles into the system by entrainment of gas in a vortex generated by a stirred paddle.

Figure 1-44 shows a cross section of a dispersed gas flotation cell that utilizes a mechanical rotor.

Rotor creates a vortex and vacuum within the vortex tube

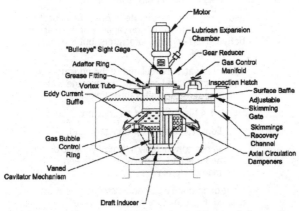

FIGURE 1-44 Cross section of a mechanical induced dispersed gas flotation unit.

Shrouds assure that the gas in the vortex mixes with and is entrained in the water

Rotor and draft inducer causes the water to flow as indicated by the arrows in this plane while also creating a swirling motion

A baffle at the top directs the froth to a skimming tray as a result of this swirling motion

Most mechanical induced units contain three or four cells.

Figure 1-45 illustrates a four-cell unit.

Bulk water moves in series from one cell to the other by underflow baffles.

Each cell contains a motor-driven paddle and associated bubble generation and distribution hardware.

Field tests have indicated that the high intensity of the mixing in each cell creates the effect of plug flow of the bulk water from one cell to the next.

Thus, there is no short-circuiting or breakthrough of a part of the inlet flow to the outlet weir box.

Mechanical complexity makes mechanical induced flotation units the most maintenance-intensive of all gas flotation configurations.

As a result of the need for motor shaft seals on penetrations to the cell, these units have

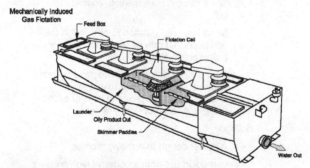

FIGURE 1-45 Cross section of a four-cell mechanical induced flotation unit.

traditionally operated very near atmospheric pressure.

Each of the above processes assumes the inlet water to the flotation unit is already at atmospheric pressure.

> When the upstream primary separator operates at elevated pressures, substantial gas saturation of the produced water may already exist.

> In these cases, flashing to atmospheric pressure may be sufficient to generate bubbles without added gas saturation.

Other Configurations

The combination of dissolved gas flotation and CPIs have been attempted, with injection of a recycled portion of the effluent from the CPI into the influent stream.

> Limited field data is available.

> Field tests have not been encouraging.

> It is not recommended because when the dissolved air breaks out of solution, turbulence that can adversely affect the action of the CPI is created.

Many types of configurations have been developed.

> Some designs incorporate complex flow patterns.

> Cells range from one to five.

> Some designs have multiple educators per cell.

> Some have recirculation rates through the educators that may be several multiples of the bulk water throughput rate.

> The concepts described above should give the engineer some guidance to add in understanding the pros and cons of each manufacture's proprietary designs.

Sparger:

> A newer design that shows promise

> Introduces gas from an external high pressure source similar to that of an aerator in an aquarium

Porous media nozzles are used to form very small bubbles, the size of which is controlled by the pore size in the media

For the most effective attachment of oil droplets to these sparged bubbles

> Bubble size should be approximately the same as the smallest oil droplets to be removed.

> Due to the small bubbles used by the sparger, long retention times (e.g., 10 min) are required.

Generates some mechanical complexity due to the needs for a separate pressurized gas supply and for numerous porous media nozzles that may be prone to plugging

Using multiple spargers could:

> Generate smaller bubbles

> Generate flow rates

> Generate better gas mixing with produced water than other designs

The detriment is that the porous media could plug with time leading to high maintenance and poor availability.

Sizing Dispersed Gas Units

An effective design requires:

> High gas induction rate

> Small-diameter induced gas bubble

> Relatively large mixing zone

The design of the nozzle or rotor, and of the internal baffles, is thus critical to the unit's efficiency.

Nozzles, rotors, and baffles are patented designs.

Field tests indicate:

> Units operate on a constant percent removal basis

> Within normal ranges their oil removal efficiency is independent of inlet concentration or oil droplet diameter

> Properly designed unit with suitable chemical treatment should have oil removal efficiency of:

40–55% per active cell

Overall of 90%

Excellently designed systems may yield an efficiency of 95%

Poorly designed, poorly operated or difficult oil-water chemistry could degrade efficiency to as low as 80%

Equation 1-25 verifies the above efficiencies. For example,

Three-cell unit can be expected to have an overall efficiency of 87%.

Four-cell unit can be expected to have an overall efficiency of 94%

Unit's actual efficiency will depend on many factors that cannot be controlled or predicted in laboratory or field tests.

Each cell is designed for approximately 1 minute retention time:

Allows gas bubbles to break free of the liquid

Forms the froth at the surface

Each manufacturer gives the dimensions of its standard units and the maximum flow rate based on this criteria.

Figure 1-46 is a typical graph showing the effluent quality versus influent quality for a four-cell dispersed gas flotation unit.

At inlet concentrations less than 200 mg/l, the oil removal efficiency declines slightly.

At low oil inlet concentrations it is more difficult for the flotation unit to achieve intimate contact and interaction between the gas bubbles and dispersed oil droplets.

Figure 1-46 may understate the effluent concentrations for influent oil concentrations less than 200 mg/l.

Depending on the oil concentration in the influent and the quality requirements of the effluent:

Flotation may or may not serve as a standalone process in produced water treating

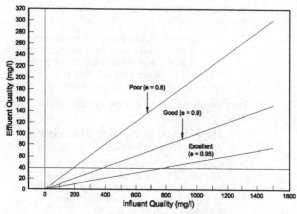

FIGURE 1-46 Effluent quality versus influent quality.

Water quality from primary production separators tend to range from 500 to 2000 ppm

From Figure 1-46, a well-designed gas flotation unit would be limited to an effluent quality in the 30 to 80 ppm range when used as the sole water treating unit downstream of the primary separator

Since separation efficiency is independent of the influent oil concentration, upsets in the primary separator operation could make a significant difference in the gas flotation effluent quality

To meet effluent concentrations in the 30 to 50 ppm range, it is usually necessary to combine a gas flotation unit with some unit between the flotation unit and the primary separator, such as a CPI

Flotation units require a customized chemical treatment program to achieve adequate results.

If produced water originates from several sources in variable quantities, the development of a chemical treatment program may be difficult.

Skimmed oily water volumes are typically 2 to 5% of the machine's rated capacity and can be as high as 10% when there is a surge of water flow into the unit.

Since skimmed fluid volume is a function of weir length exposure over time, operation of the unit at less than design capacity increases the water residence time but does not decrease the skimmed fluid volumes.

Flotation units normally include multiple cells in series.

Mechanical failure in a single cell causes a degradation in performance.

For example, a four-cell unit with a mechanical failure in one cell becomes a three-cell unit capable of separating 87.5% of the dispersed oil in the feed stream [i.e., $1 - (1 - 0.5)^3 = 0.875$].

Gas flotation equipment is typically purchased as a prefabricated unit selected from a vendor's list of standard size units, rather than being custom specified and designed for each specific application.

As a result, the bulk of the design is performed by the vendor, and relatively little design opportunity exists for the user.

A tabulation of representative sizes, weights, horsepower, and residence times for both hydraulically and mechanically induced flotation units is illustrated in Table 1-4.

Performance Considerations

Several factors that should be taken into account to maintain performance include:

Cells must be properly leveled on initial installation and this condition must be maintained. Since skimming depends on proper operation of a weir, small out-of-level conditions will prevent proper skimming of oil. Movement of the flotation cells can also set up liquid surges that can prevent proper skimming.

Liquid levels must be carefully controlled to permit proper weir operation. Level control system parameters must be carefully set to prevent liquid level oscillations. Throttling valves are preferred over snap acting valves on

Table 1-4 Characteristics of Representative Gas Flotation Units

IGF Type	Company	Brand	Model	Power HP	Length ft (ss)	Width ft (OD)	Fluid Volume ft3	Fluid Volume gal	Nominal Flow bwpd	Nominal Flow ft3min	Residence min
Hydraulic IGF	Wemco	ISF	30X	15	20.5	4.5	326	2439	10,200	39.8	8.2
			75X	30	29	5.5	689	5154	25,700	100.2	6.9
			160X	50	33	7.5	1457	10,900	54,800	213.7	6.8
	ESI	Tridar	DL-100	7.5	15.5	5	304	2274	10,000	39.0	7.8
			DL-200	20	20.5	9	433	3239	20,000	78.0	5.6
			DL-500	50	23	14	1602	11,985	50,000	195.0	8.2
	Monosep	Verisep	3MV	3	15	3.5	144	1077	3000	11.7	12.3
			10MV	8	15	6	424	3172	10,000	39.0	10.9
			50MV	25	31.3	9.5	1798	13,451	50,000	195.0	9.2
Mechanical IGF	Serck Baker	Depurator	SB-020	6	14	2.5	32	240	2000	7.8	4.1
			SB-100	8	21	4.5	209	1563	10,000	39.0	5.4
			SB-500	50	37	7	898	6714	50,000	195.0	4.6
	Petrolite		GFS-5	12	22	3.5	308	2304	5000	19.5	15.8
			GFS-10	20	27	5	540	4040	10,000	39.0	13.8
			GFS-45	40	37	6	1332	9965	45,000	175.5	7.6
	Wemco	Depurator	36	12	14.4	3.5	81	606	1720	6.7	12.1
			56	20.5	26.5	5.7	346	2588	10,300	40.2	8.6
			84X	60.5	34.1	8.9	1390	10,399	50,000	195.0	7.1
			144X	120.5	64.2	12	5832	43,629	1,714,000	668.3	8.7

both the water inlet and outlet. The flow disturbances caused by the rapid opening and closing of the snap acting valves may generate level disturbances. Gravity flow of the inlet feed stream to the gas flotation unit is preferred over pumping. The high shearing action created in a pump will break up the larger oil droplets into smaller droplets, making separation more difficult.

Many induced flotation units, particularly mechanical flotation units, operate at pressures within a few ounces of atmospheric pressure. The walls are thin and have numerous penetrations for motor shafts and observation hatches. As a result of the simplicity of design, air can easily enter the units around the paddle if observation hatches are left open. Oxygen in the water-treating system increases the corrosion rate in the unit as well as all downstream carbon steel equipment and can cause the formation of a reddish precipitate resulting from oxidation of dissolved iron in the treated water. To avoid corrosion and the precipitate, care should be taken to avoid oxygen ingress. Hatches should be left closed as much as possible, and the integrity of shaft seals should be maintained.

Proper chemical treatment is essential to the operation of gas flotation. Care must be taken to ensure that the chemical injection facilities are operating as expected and that proper dosage is administrated and mixed, both to promote sufficient separation and to prevent excessive chemical use. The customized chemical treatments involving polyelectrolytes, de-emulsifiers, scale inhibitors, and corrosion inhibitors may result in chemical incompatibilities, either between chemicals or between chemical treatments and flotation cell materials. These incompatibilities may be compounded by propagation through the treatment facility of any chemicals added upstream of the water treating system. Units should be monitored for any unexpected sludge or precipitates or for unexpectedly high corrosion or elastomer deterioration rates.

Field tests indicate the following performance findings:

> Induced gas flotation units remove almost 100% of oil droplets 10 to 20 microns and above and have some effect on oil droplets in the 2 to 5 micron range.

> Oil removal efficiency depends on choosing the correct chemical and chemical dosage (refer to Figure 1-47).

> Mechanical units tend to be more efficient than hydraulic units with influent concentrations, from 50 to 150 mg/l, while hydraulic units are more efficient above 500 mg/l. Both units are equally efficient between 150 to 500 mg/l.

> The performance of all induced gas flotation units are relatively insensitive to flow-rate variations between 70 to 125% of the design flow rate.

> Mechanically induced units appear to tolerate greater throughput rate fluctuations than hydraulic induced units.

> The separation efficiency of all units depends in the influent concentration.

> Changing the water temperature from ambient to 140°F (60°C) results in a slight improvement in oil recovery at normal pH values.

Gas flotation units should be used when:

> The inlet oil concentrations are not too high (250–500 mg/l)

> The effluent discharge requirements are not too severe (25-50 mg/l)

> Chemical companies are available to formulate an appropriate chemical treatment program

> Power costs are low or moderate

Gas flotation units should not be used when:

> Equipment size and weight are prime considerations

> The unit is subject to acceleration and tilting, such as floating production facilities

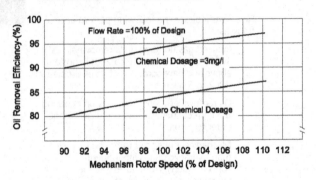

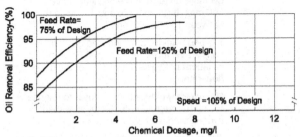

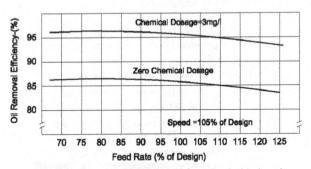

FIGURE 1-47 Oil removal efficiencies of mechanical induced flotation units.

The water stream to be treated is comprised of multiple water sources having significantly varying water chemistry and dispersed oil characteristics

Service support from water treating chemical vendors is limited

Very low effluent oil concentrations are
required

Power costs are high

Hydrocyclones

General Considerations

Hydrocyclones have been used in produced water
treating systems to de-oil the water since the early
1980s.

They are referred to as "liquid–liquid de-oiling"
hydrocyclones.

They are sometimes called "enhanced gravity
separators."

They are further classified as static or dynamic
hydrocyclones.

Operating Principles

Hydrocyclones use centrifugal force to remove oil
droplets from oily water.

Figure 1-48 shows a static de-oiling hydrocyclone
consisting of liner(s) contained within a pressure
retaining outer vessel or shell.

The liner consists of the following four sections:

Cylindrical swirl chamber

Concentric reducing section

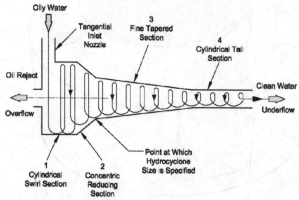

FIGURE 1-48 Liquid–liquid static hydrocyclone separation liner.

Fine tapered section

Cylindrical tail section

Figure 1-49 shows a typical multiliner vessel.

Oily water enters the cylindrical swirl chamber through a tangential inlet nozzle (Figure 1-50), creating a high velocity vortex with a reverse-flowing central core.

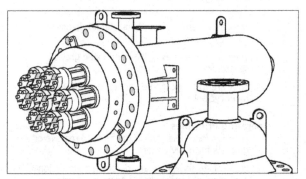

FIGURE 1-49 Multiliner hydrocyclone vessel.

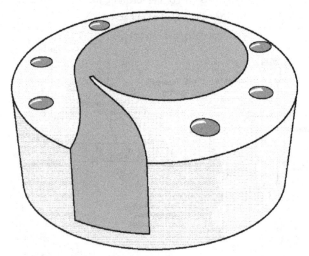

FIGURE 1-50 Tangential inlet nozzle.

The fluid accelerates as it flows through the concentric reducing section and fine tapered section.

The fluid then continues at a constant rate through the cylindrical tail section.

Larger oil droplets are separated from the fluid in a fine tapered section, while smaller droplets are removed in the tail section.

Centripetal forces cause the lighter-density droplets to move toward the low-pressure central core, where axial reverse flow occurs.

Oil is removed through a small-diameter port located in the head of the hydrocyclone and is known as the "reject stream" or "overflow."

Clean water is removed through the downstream outlet and is known as the "underflow."

Separation Mechanism

The separation mechanism is governed by Stokes' law.

It features gravitational forces in the orders of magnitude between 1000 and 2000 gs higher than that available in conventional gravity-based separation equipment.

A high-velocity vortex with a reverse-flowing central core is set up by entry through a specially designed tangential inlet(s) (refer to Figure 1-50).

Fluid is accelerated (thereby offsetting the frictional losses) through the concentric reducing and fine tapered sections of the cyclone, where the bulk of separation occurs into the cylindrical tail section where smaller, slower-moving droplets are recovered.

The size of a hydrocyclone (for example, 35 mm or 60 mm) refers to the diameter at transition between the concentric reducing and fine tapered section of the cyclone (refer to Figure 1-48).

Orientation and Operating Considerations

Hydrocyclones can be orientated either horizontally or vertically.

Horizontal orientation:

Is the most common

Requires more plan area (deck space)

Is more convenient for maintenance (about 42-inch clearance is required to remove the liners from the vessel)

The energy required to achieve separation is provided by the differential pressure across the cyclone.

A minimum of 100 psi is required.

Higher pressures are preferable when available.

Reject stream:

1 to 3% of volume percent of inlet

10% (by volume) is oil, the rest is water

May be directed back to the separator through a low shear progressive cavity pump

In some fields, oilfield chemicals:

Cause swelling of the rubber stator of these pumps, leading to poor performance

In such situations, low-speed centrifugal pumps with an open impeller may be suitable

Many installations include a degassing vessel downstream of the clean product water outlet

The vessel provides:

Short residence time serving essentially as a gas flotation unit

Oil slug-catcher volume in case of major upsets

Additional residence time for emulsion-breaking chemicals

Static Hydrocyclones

Static hydrocyclones require a minimum pressure of 100 psi so as to produce the required velocities.

If 100 psi is not available:

A low-shear progressive-cavity pump should be used

Sufficient pipe should be used between the pump and the hydrocyclone

It should allow pipe coalescence of the oil droplets

These do not appear to work well with oil droplets less than 30 microns in diameter.

Performance

The performance of static hydrocyclones is chiefly influenced by:

Reject ratio

Pressure drop ratio (PDR)

Reject ratio:

Defined as the ratio of the reject fluid rate (volume of oil and water discharged from the reject outlet) to the total inlet volume flow rate expressed as a percentage

Reject orifice size is fixed (typically 2 mm)

Controlled by the back pressure, directly proportional to the pressure drop ratio, on the reject outlet stream

Recommended reject ratio is 1 to 3% of feed flow; optimum is about 2%

2% provides a safety margin to ensure that the efficiency is not affected by upset conditions, such as:

Fluctuations in pressure drop across the unit, or

Slight surge in oil concentration which could affect efficiency

Operating below the optimum reject ratio results in low oil removal efficiencies

To maintain efficiency at a lower reject ratio:

Reject port diameter must be made smaller

Which increases the probability of it becoming blocked

Operating above the optimum reject ratio does not impair oil removal efficiencies.

Pressure drop ratio (PDR):

> Refers to the ratio of the pressure difference between

>> The inlet and reject outlets, and

>> The inlet and water outlet

> A PDR of between 1.4 and 2 is usually desired

> Performance also affected by:

>> Oil drop size

>> Concentration of inlet oil

>> Differential specific gravity

>> Inlet temperature

>> Temperatures greater than 80°F result in better operation

Performance considerations:

> Varies from facility to facility (as with flotation units)

> An assumption of 90% oil removal is reasonable for design

> Performance cannot be predicted more accurately from laboratory or field testing because it is dependent on:

>> Actual shearing and coalescing that occur under field flow conditions

>> Impurities in the water such as:

>>> Residual treating and corrosion chemicals

>>> Sand

>>> Scale

>>> Corrosion products, which vary with time

Excellent coalescing devices actually function best as a primary treating device followed by a downstream skim vessel that can separate the 500 to 1000 micron droplets that leave with the water effluent.

Figure 1-51 shows a simplified P&ID for a hydrocyclone.

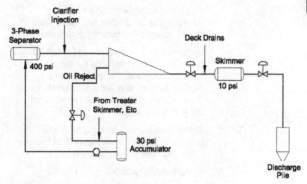

FIGURE 1-51 Simplified P&ID showing a hydrocyclone used as preliminary treating device.

Advantages of Hydrocyclones

No moving parts (thus, minimum maintenance and operator attention are required)

Compact design (reduces weight and space requirements)

Insensitive to motion (suitable for floating facilities)

Modular design (allows easy addition of capacity)

Offer lower operating costs when compared to flotation units, if inlet pressure is available

Disadvantages of Hydrocyclones

Need to install a pump if oil is available only at low pressure

Tendency of the reject port to plug with sand or scale

Sand in the produced water will cause erosion of the cones and increase operating costs

Performance of hydrocyclones is also affected by the following parameters:

Oil drop size (at fixed concentrations)

Efficiency generally decreases as the oil droplet size is reduced.

This is consistent with Stokes' law, where a smaller droplet will move less rapidly toward the hydrocyclone core.

Droplets below a certain size (about 30 microns) are not captured by the hydrocyclone and, therefore, as the median feed oil droplet size decreases, more of the smaller droplets escape and the efficiency drops.

Restrictions (valves, fittings, etc.) and pumps causing droplet shearing in the incoming flow should be avoided.

Differential specific gravity

At a constant temperature, the hydrocyclone oil removal efficiency increases as the salinity increases and/or the crude specific gravity decreases.

As the specific gravity difference between water and oil increases, a greater driving force for oil removal in the hydrocyclone occurs.

Inlet temperature

The temperature of the produced water inlet stream determines the viscosity of the oil and water phases and the density difference between the two phases.

As temperature increases, the water viscosity decreases slightly, while the density difference increases more substantially.

This is because oil density decreases at a faster rate than the water density.

Inlet flow rate

The centrifugal force induced in the hydrocyclone is a function of flow rate.

At low flow rates, insufficient inlet velocity exists to establish a vortex and separation efficiency is low.

Once the vortex is established, the efficiency increases slowly as a function of the flow rate to a point where the pressure at the core approaches atmospheric.

Any further increase in the flow rate hinders oil flow from the reject outlet and causes efficiency to decline.

In addition, a high flow rate can cause shearing of the droplets.

This maximum flow rate is the "capacity" of the chamber.

Flow rate is controlled by back pressure on the underflow outlet.

The ratio of maximum to minimum flow rate, as determined by the lowest separation efficiency acceptable and the available pressure drop, is the "turndown ratio" for a given application.

Dynamic Hydrocyclones

A dynamic hydrocyclone uses an external motor to rotate the outer shell of the hydrocyclone, whereas in a static hydrocyclone the outer shell is stationary and feed pressure supplies the energy to accomplish separation of oil from water (no external motor required).

As shown in Figure 1-52, it consists of the following components:

> Rotating cylinder
>
> Axial inlet and outlet
>
> Reject nozzle
>
> External motor

Rotation of the cylinder creates a "free vortex."

The tangential speed is inversely proportional to the distance to the centerline of the cyclone.

Since there is no complex geometry that requires a high pressure drop, dynamic units can operate at lower inlet pressures (approximately 50 psig) than static units.

The effect of the reject ratio is not as important in dynamic units as it is in static units.

Dynamic hydrocyclones have found few applications due to poor cost–benefit ratio.

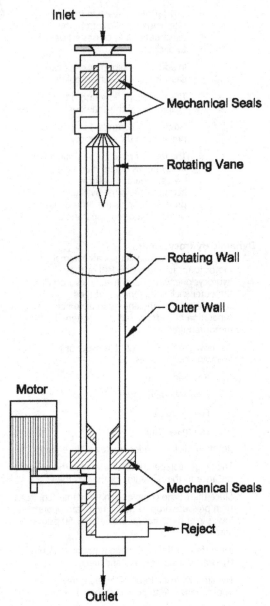

FIGURE 1-52 Liquid–liquid dynamic hydrocyclone separation.

Selection Criteria and Application Guidelines

Hydrocyclones can be used when:

> Median oil particle size is in excess of 30 microns
>
> Produced water feed pressure is at least 100 psig
>
> Platform deck space is a critical consideration
>
>> Hydrocyclones have low size and weight requirements compared to other water treating equipment for the same capacity.
>
> An appreciable quantity of solids is not present
>
> An appreciable amount of free gas is not present
>
> The flow rate and feed water oil concentrations are fairly constant
>
> Low equipment maintenance is desired; since a hydrocyclone has no moving parts, its maintenance requirements are fairly low
>
> Power constraints exist; do not require any outside energy supply, except for a low HP (about 5 HP) reject-recycle pump

They are not applicable when:

> A tight emulsion exists, with a median oil droplet size less than 30 microns (manufacturers claim that newer high-efficiency liners are capable of removing 20 microns)
>
> The feed water pressure is less than 100 psig; a pump would be required to develop adequate pressure to use the hydrocyclone (pumps cause the oil droplets to shear, making it more difficult for separation by a hydrocyclone)
>
> The difference in specific gravity between the oil and water is relatively low, that is, heavy crude is being produced
>
> Considerable sand is entrained in the produced water; the sand could potentially plug the reject orifice and also cause erosion of the liner

Sizing and Design

The performance of hydrocyclones is measured in terms of oil removal efficiency (E).

Product oil removal efficiency $(E) = \dfrac{(C_i - C_o)(100)}{C_i}$

where

C_i = dispersed oil concentration in feed water

C_o = dispersed oil concentration in effluent water

Figure 1-53 shows generalized removal efficiency curves of a hydrocyclone.

For a typical case (30° API oil and 1.05 SG water), the differential specific gravity is 0.17 and the removal efficiency would be 92% of 40 micron, 85% of 30 micron, and 68% of 20 micron oil droplets.

Figure 1-54 shows a typical control scheme for a hydrocyclone.

Disposal Piles

General Considerations

Construction:

Large-diameter (24 to 48 inches) open-ended pipe

Attached to platform and extends below the water line

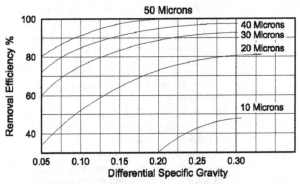

FIGURE 1-53 Generalized performance curves for a hydrocyclone.

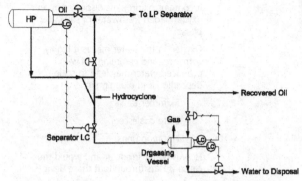

FIGURE 1-54 Typical control scheme for a hydrocyclone.

Main uses:

> Concentrate all platform discharges into one location
>
> Provide a conduit, protected from wave action, for deep discharge to prevent sheens from occurring during an upset condition
>
> Provide an alarm/shutdown point in the event of a failure that causes oil to flow overboard

Authorities having jurisdiction:

> Require all produced water to be treated (skimmer tank, coalesced, or flotation) prior to disposal in a disposal pile
>
> In some locations disposal piles are permitted to collect:
>
> > Treated produced water
> >
> > Treated sand
> >
> > Liquids from drip pans and deck drains, and
> >
> > As a final trap for hydrocarbon liquids in the event of equipment upset.

Disposal piles are excellent for deck drainage disposal.

> Deck drainage:
>
> > Flow originates either from rainwater or wash-down water.

It contains oil droplets dispersed in an oxygen-laden freshwater or saltwater phase.

Oxygen in the water makes it highly corrosive, and commingling with produced water may lead to scale deposition and plugging in

> Skimmer tanks
>
> Plate coalescers
>
> Flotation units.

Flow is highly irregular and would thus cause upsets throughout these devices.

Flow must gravitate to a low point for collection and is either pumped up to a higher level for treatment or treated at that low point.

Disposal piles:

> Can be protected from corrosion
>
> By design located low enough on the platform to eliminate the need for pumping the water
>
> Not severely affected by large instantaneous flow-rate changes (effluent quality may be affected to some extent, but the operation of the pile can continue)
>
> They contain no small passages subject to plugging by scale build-up
>
> They minimize commingling with the process since they are the last piece of treating equipment before disposal

Disposal Pile Sizing

Produced water treating equipment can treat smaller oil droplets than those that can be predicted to settle out a slender disposal pile.

Small amounts of separation will occur due to coalescence in the inlet piping and pipe itself.

However, no significant treating of produced water can be expected.

If the deck drainage is merely contaminated rainwater, the disposal pile diameter can be estimated from the following equation, assuming the need to separate 150-micron droplets:

Field units

$$d^2 = \frac{0.3(Q_W + 0.356 A_D R_W + Q_{WD})}{(\Delta SG)} \quad (1\text{-}26a)$$

SI units

$$d^2 = \frac{28,289(Q_W + 0.356 A_D R_W + Q_{WD})}{(\Delta SG)} \quad (1\text{-}26b)$$

where

d = pipe internal diameter, in. (mm)

Q_W = produced water rate (if in disposal pipe) bwpd (m³/hr)

A_D = plan area of deck, ft.² (m²)

R_W = rain fall rate, in./hr (mm/hr) = 2 in./hr (50 mm/hr) normally assumed

ΔSG = difference in SG between oil droplets and rain water

Q_{WD} = wash-down rate, bpd (m³/hr) = 1500N (9.92 N)

N = number of 50 gpm (189.25) wash-down hoses

Comments on Equation 1-26:

Either the wash-down rate or the rainfall rate should be used as it is highly unlikely that both would occur at the same time.

The produced water rate is only used if the produced water is routed to the pile for disposal.

Disposal pile length:

Shallow waters

Should be as long as the water depth permits to:

Provide for maximum oil containment in the event of an upset

Minimize the appearance of any sheen

Deep waters

Length is set to assure that:

An alarm and then a shutdown signal can be measured before the pile fills with oil

Signals must be high enough so as not to register tide changes

Disposal Piles (Offshore Platforms)

Length of the pile submergence below the normal water level required to assure that a high level will be sensed before the oil comes within 10 ft. (3 m) of the bottom is given by:

Field units

$$L = \frac{(H_T + H_S + H_A + H_{SD})(SG)}{(\Delta SG)} + 10 \qquad (1\text{-}27a)$$

SI units

$$L = \frac{(H_T + H_S + H_A + H_{SD})(SG)}{(\Delta SG)} + 0.6 \qquad (1\text{-}27b)$$

where

L = depth of pile below MWL, ft. (m)

H_T = normal tide range, ft. (m)

H_s = design annual storm surge, ft. (m)

H_A = alarm level (usually 2 ft. (0.6 m)), ft. (m)

H_{SD} = shutdown level (usually 2 ft. (0.6 m)), ft. (m)

SG = specific gravity of the oil relative to water

It is possible in shallow water to measure the oil–water interface for alarm or shutdown with a bubble arrangement and a shorter pipe.

This is not recommended where water depth permits a longer pile.

To minimize wave action effects a minimum pile length of 50 feet is required.

Figure 1-55 is a schematic showing disposal pile length.

Skim Piles

General Considerations

A skim pile adds a series of inclined coalescing plates with oil collection risers.

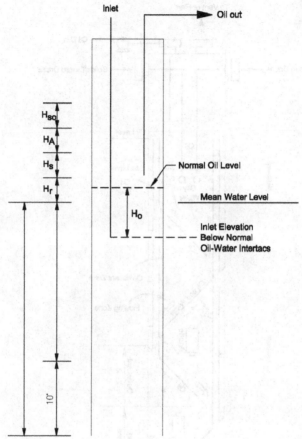

FIGURE 1-55 Disposal pile length.

Operation

As shown in Figure 1-56:

Flow through the multiple series of baffle plates creates zones of no flow that reduce the distance a given oil droplet must rise to be separated from the main flow.

Once in this zone, there is plenty of time for coalescence and gravity separation.

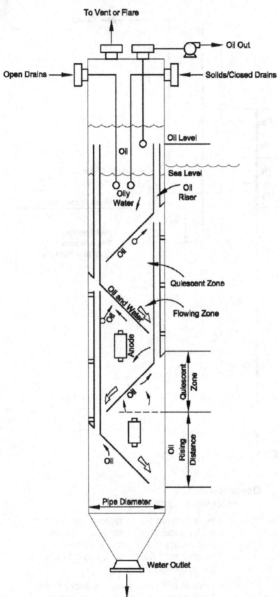

FIGURE 1-56 Cross section showing flow pattern of a skim pile.

The larger droplets then migrate up the underside of the baffle to an oil collection system.

Advantages

More efficient at separating oil from water

Provides for some degree of sand cleaning

Regulatory Considerations

Most authorities having jurisdiction state that produced sand must be disposed of without "free oil."

It is doubtful that sand from a vessel drain meets this criteria when disposed of in a standard disposal pile.

Sand traveling the length of the skim pile will abrade on the baffles and be water washed. This removes the "free oil" that is then captured in a quiescent zone

Skim Pile Sizing

Determination of skim pile length is the same as that for any other disposal pile.

Because of the complex flow regime, a suitable equation has yet to be developed to size skim piles for deck drainage.

Field experience has indicated that acceptable effluent is obtained with 20 minutes' retention time in the baffled section of the pile.

Using this and assuming that 25% of the volume is taken up by the coalescing zones, we have the following:

Field units

$$d^2 L' = 19.1(Q_w + 0.356 A_D R_W + Q_{Wd}) \quad (1\text{-}28a)$$

SI units

$$d^2 L' = 565,811(Q_w + 0.001 A_D R_W + Q_{Wd})$$
$$(1\text{-}28b)$$

where

L' = Length of baffle section, ft. (m) [the submerged length is $L' + 15$ ft. ($L' + 4.6$ m) to allow for an inlet and exit from the baffle section]

DRAIN SYSTEMS

Pressure (Closed) Drain System

The pressure (closed) drain system is connected directly to pressure vessels.

Liquid contains dissolved gases that flash in the drain system and can become a hazard if not handled properly.

The drain valve could be accidently left open.

Once liquid drains out of the vessel a large amount gas will flow out of the vessel into the closed drain system (gas blow-by) and will have to be handled safely.

It should always be routed to a pressure vessel and should never be connected to an open drain system.

Atmospheric (Open) Drain System

An atmospheric (open) drain system collects liquids that spill on the ground.

It is also called "gravity" drain.

Liquids gathered in an open drain system is typically rainwater or wash-down water contaminated with oil.

The oil is usually circulated back into the process. Every attempt should be made to minimize the amount of aerated water that is recycled with the oil.

Oxygen in the water:

Increases corrosion

Commingled with produced water may lead to scale deposition and plugging

This is best achieved by routing open drains to a sump tank that has a gas blanket and operates as a skimmer.

To keep gas from the skimmer from flowing out of the drain, a water seal should be built into the inlet to the sump tank.

Water seals should also be installed on laterals from separate buildings or enclosures to keep the open drain system from being a conduit of gas from one location in the facility to another.

Environmental Considerations

Authorities having jurisdiction require all produced water to be treated (skimmer tanks, coalescers, and flotation units) prior to disposal in a disposal pile.

Sample of produced water effluent limitations:

Angola: 30 mg/l all facilities

Brunei: 30 mg/l all facilities

Ecuador: 30 mg/l all facilities

Indonesia: 30 mg/l new facilities, 40 mg/l "grand fathered" facilities

Malaysia: 40 mg/l all facilities

Nigeria: 30 mg/l all facilities

North Sea: 30 mg/l all facilities

Thailand: 40 mg/l all facilities

U.S.A.: mg/l average OCS waters, 0 discharge inland waters

Disposal piles are permitted to collect treated produced water, treated sand, liquids from drip pans and deck drains, and as a final trap for hydrocarbon liquids in the event of equipment upsets.

INFORMATION REQUIRED FOR DESIGN

Design Basis

The first step in designing a water-treating system is to establish the design basis.

To establish this basis, a variety of information must be obtained, including the type and location of the facility and the method of water disposal to be used.

Effluent Quality

The method of disposal and the facility location are the primary factors in determining the effluent quality requirements. For example:

If the water is to be injected into a disposal well, then the permeability of the injection formation may determine the effluent oil quality.

If solids must be filtered, it may be necessary to treat the water to a low level of oil content (25 to 50 mg/l) to keep from plugging the filter.

If the facility is located offshore and the water is to be disposed of overboard, then the location of the platform determines the effluent quality.

The Environmental Protection Agency (EPA) establishes the maximum amount of oil and grease that may be discharged with produced water into navigable waters of the United States.

In state waters, the state or local governments establish the water discharge criteria.

The EPA requirements for the Gulf of Mexico currently state that no free oil may be discharged with produced sand or with deck drainage.

The current produced water limit is 29 mg/l of oil determined from an average of four samples taken within a 24-hour period once per month.

These limits are subject to change, and the requirements for each location should be researched before a system is designed.

Examples of worldwide produced water effluent oil concentration limitations include:

Angola: 30 mg/l all facilities

Brunei: 30 mg/l all facilities

Ecuador, Colombia, Brazil, Argentina: 30 mg/l all facilities

Indonesia: 15 mg/l new facilities, 25 mg/l "grand fathered" facilities

Malaysia, Middle East: 30 mg/l all facilities

Nigeria, Cameroon, Ivory Coast: 40 mg/l all facilities

North Sea, Australia: 30 mg/l all facilities

Thailand: 30 mg/l all facilities

U.S.A.: 29 mg/l average OCS waters, 0 discharge inland waters

Produced Water Flow Rate

The flow rate of water that a treating system and each component of a system must handle must be know in order for a treating system to be designed.

This water flow rate may take into account produced water, rainwater, washdown water, and/or fire deluge water.

These sources of water may not all be present at a given facility, and even if they are, they may not all be additive.

That is, the rainwater and the washdown water, for example, need not be treated simultaneously.

Only the larger of these two flow rates need be considered.

A typical design rainfall rate for the Gulf of Mexico is 2 in./hr (50 mm/hr) applied over the platform deck area.

For washdown water, a rate of 1500 bwpd per washdown hose may be used. Fire deluge systems typically distribute 0.25 gal/min ft^2 (0.95 l/min m^2 water over the platform deck area.

Water Specific Gravity

The specific gravity of the water is important in the design of equipment based on gravity separation.

If this data is not available, a value of 1.07 may be assumed for produced water.

Water Viscosity

The water viscosity at the treating temperature affects the design of gravity settling equipment.

If data is not available, a value of 1.0 cp may be assumed.

Oil Concentration

One must determine the concentration of the oil in the influent produced water.

This is best determined from samples and laboratory data.

Various attempts have been made to develop procedures to determine the oil concentration in the water outlet from free-water knockout and heater treaters.

These attempts have not been very successful.

The oil concentration can vary over a wide range, but a conservative assumption of 1000 to 2000 mg/l may be used for properly designed equipment.

It is possible theoretically to trace the oil droplet size distribution through the tubing, choke, flowlines, separators, dump valves, and other equipment.

When the distribution of oil droplets entering the free-water knockout is known then the theoretical separation in the free-water knockout may be used to calculate the outlet oil concentration; however, many of the parameters needed to solve these equations, especially those involving coalescence, are unknown.

Soluble Oil Concentration

The soluble oil concentration at the discharge conditions should be determined.

Conventional water treating equipment does not remove soluble oil from the water.

The EPA, however, does not differentiate between soluble and dispersed oil.

Thus, the soluble oil concentration must be subtracted from the discharge limit to indicate the maximum amount of dispersed oil allowed in the effluent.

Typical Gulf of Mexico values for soluble oil concentration range from 0 to 30 mg/l, although readings as high as 100 mg/l have been recorded.

If measurements are not available, an allowance of 15 mg/l for dissolved oil is recommended.

Oil Specific Gravity

One must determine the specific gravity of the oil at treating conditions in order to size gravity settling equipment.

Oil Droplet Size Distribution

The water level in the free-water knockout and/or heater treaters upstream of the treating system is usually controlled by a water dump valve.

Because of the dispersion through the water dump valve, the oil drop size distribution at the outlet of a free-water knockout or a heater treater is not a significant design parameter.

Equations 1-29 and 1-30 can be used to derive a drop size after the fluid passes through the dump valve.

$$d_{max} = \frac{0.725}{\varepsilon^{2/5}} \left[\frac{\sigma}{\rho_w} \right]^{3/5} (10)^4 \qquad (1-29)$$

where

d_{max} = diameter of droplet above which size only 5% of the oil volume is contained, in microns

ε = mixing parameter equivalent to the work done on a fluid per unit mass per unit time, in cm^2/s^2

σ = surface tension, in dynes/cm

ρ_w = water density, in gram/cm^3

$$\varepsilon = 1150 \frac{\Delta p}{t_r} \qquad (1-30)$$

For a pressure drop of 50 to 75 psi (340 to 520 kPa), one would expect a maximum droplet of 10 to 50 microns (μ), independent of the droplet size distribution upstream of this valve.

If there were sufficient time as defined by:

$$t = \frac{\pi}{6} \left[\frac{d^j - (d_o)^j}{\varphi K_s} \right] \qquad (1-31)$$

for coalescence to occur in the piping downstream of the dump valve, then all the practices would coalesce to the diameter defined by Equations 1-29, 1-30, and 1-31 for the pressure drop in the piping.

It is extremely difficult to evaluate Equations 1-29, 1-30, and 1-31.

Therefore, it is recommended that a maximum diameter of 250 to 500 microns (μ) be used for design if no other data is available. (*Note*: 250 microns is the more conservative assumption.)

It is clear that there will be a distribution of droplet sizes from zero to the maximum size, and this distribution will depend on parameters unknown at the time of initial design.

Experimental data indicates that the conservative assumption for design would be to characterize the distribution as a straight line, as shown in Figure 1-3.

Oil Drop Size Distribution: Open Drains

EPA regulations require that "free oil" be removed from deck drainage prior to disposal.

It is extremely difficult to predict an oil drop size distribution for rainwater or washdown water that is collected in an open drain system, and regulations do not define what size droplet is meant by "free oil."

It is long-standing refinery practice to size the drain water treating equipment to remove all oil droplets 150 microns (μ) in diameter or larger.

If no other data is available, it is recommended that this be used in sizing sumps and disposal piles that must treat deck drainage.

Equipment Selection Procedure

It is desirable to bring information included earlier into a format that can be used by the design engineer in selecting and sizing the individual pieces of equipment needed for a total water treating system.

Authorities having jurisdiction require that produced water from the free-water knockout receive at least some form of primary treatment before being sent to a disposal pile or skim pile.

Deck drainage may be routed to a properly sized disposal pile that will remove "free oil."

Every water-treating system design must begin with the sizing, for liquid separation of a free-water knockout, heater treater, or three-phase separator. These vessels should be sized in accordance with the procedures discussed in the appropriate design guidelines.

With the exception of these restraints the design engineer is free to arrange the system as he or she sees fit.

There are many potential combinations of the equipment previously described.

Under a certain set of circumstances, it may be appropriate to dump the water from a free-water knockout directly to a skim tank for final treatment before discharge.

Under other circumstances a full system of plate coalescers, flotation units, and skim piles may be needed.

In the final analysis, the choice of a particular combination of equipment and its sizing must rely rather heavily on the judgment and experience of the design engineer.

The following procedure is meant only as a guideline and not as a substitute for this judgment and experience.

Many of the correlations presented herein should be refined as new data and operating experience become available. In no instance is this procedure meant to be used without proper weight given to operational experience in the specific area.

1. Determine the oil content of the produced water influent. In the absence of other information, 1000 to 2000 mg/l could be assumed.

2. Determine the dispersed oil effluent quality. In the absence of other information, use 14 mg/l for design in the Gulf of Mexico and other similar areas (29 mg/l allowed less 15 mg/l dissolved oil).

3. Determine oil drop size distribution in the influent produced water stream. Use a straight-line distribution with a maximum diameter of 250 to 500 microns in the absence of better data.

4. Determine the oil particle diameter that must be treated to meet effluent quality required. This can be calculated as effluent quality divided by influent quality times the maximum oil particle diameter calculated in step 3 ($d_r = d_{max} \, C_o / C_i$)

5. Several factors affect the selection of water-treating equipment. Perhaps the main factor in equipment selection is the value of d_r calculated in step 4. If d_r is less than 30 to 50 microns, then a flotation unit or a hydrocyclone will probably be required. Proceed to step 6. If a large area is available (as in an onshore location), consider a skimmer or plate separator. Proceed to step 10.

6. Determine flotation unit's required influent quality, assuming 90% removal. The required influent quality is the required effluent quality multiplied by 10.

7. If required flotation unit influent quality is less than quality determined in step 1, determine the particle diameter that must be treated in skim tank or plate coalescers to meet this quality. This can be calculated as the flotation cell influent quality divided by the influent quality determined

in step 1 times the maximum particle diameter calculated in step 3.

8. Determine effluent from hydrocyclone, assuming that it is 90% efficient, and determine the particle diameter that must be treated in the downstream skim vessel, assuming that $d_{max} = 500$. This value can be calculated as 500 times the dispersed oil effluent quality (step 1) divided by the effluent concentration from the hydrocyclone.

9. In sizing a skimmer, one must make several choices. First, depending on the system, a pressure vessel or an atmospheric tank should be chosen. Next, a configuration needs to be selected. Then the skimmer may be sized (refer to the appropriate equation).

10. In designing an SP Pack system, one must select the number of stages. Next, the droplet diameter which must be removed in each stage may be calculated as follows:

$$d_r = 1000 \left[\frac{C_o}{C_i} \right]^{1/n}$$

where:

d_r = oil droplet diameter to be removed in each stage, in microns

C_i = system influent oil concentration

C_o = system effluent oil concentration

N = number of stages

Once the diameter to be removed in each stage has been determined, a skimmer or plate coalesce may be selected and sized. (*Note*: the minimum diameter for a skim tank with a SP Pack mounted internally is 8 ft.)

11. Determine plate coalescers' dimensions.

   a. Choose CPI or cross-flow configuration.

   b. Determine size. Refer to appropriate equations.

12. Choose skim tank, SP Pack, or plate coalescers for application, considering cost and space available.

13. Choose method of handling deck drainage.

   a. Determine whether rainwater rate or wash-down rate governs design.

     b. Size disposal pile for drainage assuming 150-micron drop removal. (Refer to Equations 1-26 and 1-27). If disposal pile diameter is too large (greater than 48 in. (1220 mm)), consideration should be given to a skim pile. Otherwise, consider using a sump tank and a disposal pile (Refer to Equation 1-28).

14. The final equipment selection should be based on economics, which may require that the procedure be repeated to investigate alternative equipment selections. After all options have been investigated, equipment selections may be finalized and specifications prepared.

EQUIPMENT SPECIFICATION

When the equipment types are selected using the above procedure and the discharge water requirements are met, the main size parameters for each of the equipment types can be extracted from the equations presented earlier.

Skim Tank

1. Horizontal vessel designs: The internal diameter and seam-to-seam length of the vessel can be determined. The effective length of the vessel can be assumed to be 75% of the seam-to-seam length.

2. Vertical vessel designs: The internal diameter and height of the water column can be determined. The vessel height can be determined by adding approximately 3 feet to the water column height.

SP Pack System

The number and size of tanks can be determined. Alternatively, the dimensions and number of compartments in a horizontal flume can be specified.

CPI Separator

The number of plate packs can be determined.

Cross-Flow Devices

The acceptable dimensions of the plate pack area can be determined. The actual dimensions depend on the manufacturers' standard sizes.

Flotation Cell

Information is given to select a size from the manufacturers' data.

Disposal Pile

The internal diameter and length can be determined. For a skim pile the length of the baffle section can be chosen.

Example 1-2: Design the Produced Water Treating System

Given:

40° API

5000 bwpd (33 m³/hr)

Deck size is 2500 ft.² (232.3 m²)

48 mg/l discharge criteria (48 mg/l)

Actual soluble oil is 6 mg/l

Water gravity-feeds to system

Step 1. Assume oil concentration in produced water is 1000 mg/l.

Step 2. Effluent quality required is 48 mg/l and 6 mg/l dissolved oil. Therefore, effluent quality required is 42 mg/l.

Step 3. Assume maximum diameter of oil particle (d_{max}) = 500 microns.

Step 4. Using Figure 1-3, the size of oil droplet that must be removed to reduce the oil concentration from 1000 mg/l to 42 mg/l is

$$\frac{d_m}{500} = \frac{42}{1000}$$

$d_m = 21$ microns

Step 5. Consider an SP series tank treating system. See step 10. If SP Packs are not used, since

$d_m < 0$ microns, a flotation unit or hydrocyclone must be used. Proceed to step 6. (*Note*: since d_m is close to 30 microns, it may be possible to treat this water without a flotation unit. We will take the more conservative case for this example.)

Step 6. Since the flotation cell is 90% efficient, in order to meet the design requirements of 42 mg/l it will be necessary to have an influent quality of 420 mg/l. This is lower than the 1000-mg/l concentration in the produced water assumed to in step 1. Therefore, it is necessary to install a primary treating device upstream of the flotation unit.

Step 7. Using Figure 1-3, the size of oil droplet that must be removed to reduce the oil concentration from 1000 mg/l to 420 mg/l is

$$\frac{d_m}{500} = \frac{420}{1000}$$

$d_m = 210$ microns

Step 8. Inlet to water treating system is at too low a pressure for a hydrocyclone. Size a skim vessel upstream of the flotation unit.

Step 9. Skim vessel design. Pressure vessel is needed for process considerations (e.g., fluid flow, gas blow-by),

a.   Assume horizontal pressure vessel.

Settling equation

Field units

$$dL_{eff} = \frac{1000 Q_w \mu_w}{(\Delta SG)(d_m)^2}$$

$\mu_w = 1.0$ (assumed)

$(SG)_w = 1.07$ (assumed)

$(SG)_o = 0.83$ (calculated)

$$dL_{eff} \frac{(1000)(5000)(1.0)}{(0.24)(210)^2} = 472$$

Assume various diameters (d) and solve for L_{eff}.

d (in.)	L_{eff} [ft. (m)]	Actual Length [ft. (m)]
24	19.7	26.3
48	9.8	13.1
60	7.9	10.5

Retention time equation

Assume retention time of 10 minutes

$$d^2 L_{eff} = 1.4(t_r)_w\, Q_w$$

$$d^2 L_{eff} = (1.4)(10)(5000) = 7000$$

d (in.)	L_{eff} [ft. (m)]	Actual Length [ft. (m)]
48	30.4	40.4
72	13.5	17.9
84	9.9	13.1
96	7.6	10.1

SI units

$$dL_{eff} = \frac{1,145,734 Q_w \mu_w}{(\Delta SG)(d_m)^2}$$

$$\mu_w = 1.0 \text{ (assumed)}$$

$$(SG)_w = 1.07 \text{ (assumed)}$$

$$(SG)_o = 0.83 \text{ (calculated)}$$

$$dL_{eff} = \frac{(1,145,734)(33)(1.0)}{(0.24)(210)^2} = 3,572$$

Assume various diameters (d) and solve for L_{eff}.

(d) (mm)	L_{eff} (m)	Actual Length (m)
609.6	5.9	8.0
1219.2	3.0	4.0
1542	2.3	3.2

Retention time equation

Assume retention time of 10 minutes

$$d^2 L_{eff} = 42,441(t_r)_w\, Q_w$$

$$d^2 L_{eff} = (42,441)(10)(33)$$

d (mm)	L_{eff} (m)	Actual Length (m)
1219	9.4	12.3
1829	4.2	5.5
2134	3.1	4.0
2438	2.4	3.1

b. Assume vertical pressure vessel.

Field units

Settling equation

$$d^2 = 6,691F \frac{Q_w \mu_w}{(\Delta SG)(d_m)^2}$$

$F = 1.0$ assumed

$$d^2 = \frac{(6,691)(1.0)(5,000)(1.0)}{(0.24)(210)^2}$$

$d = 56.22$ in.

Retention time equation

$$H = 0.7 \frac{(t_r)_w Q_w}{d^2} \approx 0.7 \frac{(10)(5,000)}{d^2}$$

d (in.)	L_{eff} (ft.)	Seam-to-Seam Height (ft.)
60	9.72	12.7
66	8.03	11.0
72	6.75	9.8

SI units

Settling equation

$$d^2 = 6356 \times 10^8 \frac{Q_w \mu_w}{(\Delta SG)(d_m)^2}$$

$F = 1.0$ assumed

$$d^2 = \frac{(6356 \times 10^8)(1.0)(33)(1.0)}{(0.24)(210)^2}$$

$d = 1409$ mm

Retention time equation

$$H = \frac{21,218(t_r)_w Q_w}{d^2} \approx 21,218 \frac{(10)(33)}{d^2}$$

d (mm)	L_{eff} (m)	Seam-to-Seam Height (m)
1524	3.0	3.9
1676.4	2.5	3.4
1829	2.1	3.0

A vertical vessel 60 in. (1524 mm) × 12.5 ft. (3.8 m) or 72 in. (1829 mm) × 10 ft. (3 m) would satisfy all the parameters. Depending on cost and space considerations, we recommend a 72-in. (1829-mm) × 10-ft. (3-m) vertical skimmer vessel for this application.

Step 10. Investigate SP Packs in tanks as an option. Calculate overall efficiency required:

Field units

$$E_t = \frac{1000 - 42}{1000} = 0.958$$

Assume 10-ft. (3-m) diameter vertical tanks:

$$d_m^2 = 6,691 F \frac{Q_w \mu_w}{(\Delta SG)(d^2)}$$

$$d_m^2 = \frac{6,691(2)(5,000)(1.0)}{(0.24)(120)^2}$$

$$d_m = 139$$

Assume SP Pack grows 1000-micron drops:

$$E = 1 - \frac{d_m}{1,000}$$

$$= 0.891$$

One acceptable choice is two 10-ft. (3-m) diameter SP tanks in series.

$$E_t = 1 - (1 - 0.861)^2 = 0.981$$

SI units

$$E_t = \frac{1,000 - 42}{1,000} = 0.958$$

Assume 10-ft. (3-m) diameter vertical tanks:

$$d_m^2 = 6,365 \times 10^8 F \frac{Q_w \mu_w}{(\Delta SG)(d^2)}$$

$$d_m^2 = \frac{6,365 \times 10^8 (2)(33)(1.0)}{(0.24)(3,000)^2}$$

$$d_m = 139 \text{ microns}$$

Assume SP Pack grows 1000-micron drops:

$$E = 1 - \frac{d_m}{1,000}$$

$$= 0.891$$

One acceptable choice is two 10-ft. (3-m) diameter SP tanks in series.

$$E_t = 1 - (1 - 0.861)^2 = 0.981$$

Step 11. Check for alternate selection of CPI.

Field units

$$\text{number of packs} = 0.07 \frac{Q_w \mu_w}{(\Delta SG) d_m^2}$$

$$= \frac{(0.077)(5,000)(1.0)}{(0.24)(210)^2}$$

$$= 0.04 \, \text{packs,}$$

$$Q_w < 20,000 : \text{use 1 pack CPI.}$$

SI units

$$\text{number of packs} = 11.6 \frac{Q_w \mu_w}{(\Delta SG) d_m^2}$$

$$= \frac{(11.67)(33)(1.0)}{(0.24)(210)^2}$$

$$= 0.04 \, \text{packs,}$$

$$Q_w < 132 : \text{use 1 pack CPI.}$$

Step 12. Recommended skimmer vessel over CPI as skimmer will take up about the same space, will cost less, and will not be susceptible to plugging. Note that it would also be possible to investigate other configurations such as skim vessel, SP Pack, CPI, and so on, as alternatives to the use of a flotation unit.

Step 13. Sump design. Sump is to be designed to handle the maximum of either rainwater or wash-down hose rate.

a.   Rainwater rate:

Field units

Assume:

R_w = rainfall rate; 2in./hr

A_D = deck area; 2500ft.^{2}

$Q_w = 0.356 A_D R_W$

$= (0.356)(2500)(2)$

$= 1780 \text{bwpd}$

SI units

Assume

$R_W = 50.8\,\text{mm/hr}$

$A_D = 232.3\,\text{m}^2$

$Q_w = 0.001 A_D R_W$

$\quad = (0.001)(232.3)(50.8)$

$\quad = 118 \text{ m3/hr}$

b.   Wash-down rate:

Field units

Assume

$N = 2$

$Q_{WD} = 1500\,N$

$\quad = 1500(2)$

$\quad = 3000 \text{ bwpd.}$

Assume

$N = 2$

$Q_{WD} = 9.92\,N$

$\quad = 9.92(2)$

$\quad = 19.84\,\text{m}^3/\text{hr}$

The minimum design usually calls for two hoses. Because freshwater enters the sump via the drains, the sump tank must be sized using a specific gravity of 1.0 and a viscosity of 1.0 for freshwater.

c.   Assume horizontal rectangular cross-section sump.

Settling equation:

Field units

$$WL_{\text{eff}} = 70\,\frac{Q_w \mu_w}{(\Delta SG)d_m^2},$$

$$WL_{\text{eff}} = 70\,\frac{(3,000)(1.0)}{(0.150)(150)^2},$$

$$WL_{\text{eff}} = 62.2,$$

$\quad W = \text{width, ft (m)},$

$\quad L_{\text{eff}} = \text{effective length in which separation occurs, ft (m)},$

$\quad H = \text{height of tank, which is 1.5 times higher than water level within tank, or } 0.75W.$

Tank Width (ft.)	Tank L_{eff} (ft.)	Seam-to-Seam Length	Height (ft.)
4	15.6	20.0	3.0
5	12.4	16.2	3.8
6	10.4	13.5	4.5

SI units

$$WL_{eff} = 950 \frac{Q_w \mu_w}{(\Delta SG)d_m^2},$$

$$WL_{eff} = 950 \frac{(19.84)(1.0)}{(0.150)(150)^2},$$

$$WL_{eff} = 5.6,$$

W = width, ft (m)

L_{eff} = effective length in which separation occurs, ft (m),

H = height of tank, which is 1.5 times higher than water level within tank, or 0.75 W.

Tank Width (m)	Tank L_{eff} (m)	Seam-to-Seam (1.2 L_{eff})	Height (m)
1.2	4.7	6.2	0.9
1.5	3.7	4.9	1.1
1.8	3.1	4.1	1.4

A horizontal tank 6 ft. (1.83 m) × 14 ft. (4.3 m) × 5 ft. (1.52 m) would satisfy all design parameters.

d. If it is determined that the dimensions of the sump tank are inappropriately large for the platform, an SP Pack can be added upstream of the sump tank to increase oil droplet size by approximately two times the inlet droplet size.

Therefore, the sump tank size with an SP Pack can be determined by:

Field units

$$WL_{eff} = \frac{70(3,000)(1.0)}{(0.15)(300)^2}$$

$$= 15.6,$$

W = width, ft (m),

L_{eff} = effective length in which separation occurs, ft.(m),

H = height of tank, which is 1.5 times higher than water level within tank, or 0.75 W.

Tank Width (ft.)	Tank L_{eff} (ft.)	Seam-to-Seam Length	Height (ft.)
3	5.2	6.7	2.3
4	3.9	5.1	3.0
5	3.1	4.0	3.8

SI units

$$WL_{eff} = \frac{950(19.84)(1.0)}{(0.15)(300)^2}$$

$$= 1.4,$$

$W = $ width, ft (m)

$L_{eff} = $ effective length in which separation occurs, ft (m),

$H = $ height of tank, which is 1.5 times higher than water level within tank, or 0.75 W.

Tank Width (m)	Tank L_{eff} (m)	Seam-to-Seam (1.2 L_{eff})	Height (m)
0.9	1.56	2.0	0.68
1.2	1.2	1.6	0.9
1.5	0.93	1.2	1.13

A horizontal tank (with an SP Pack) 4 ft. (1.2 m) × 4 ft. (1.2 m) × 5 ft. (1.5 m) would satisfy all design parameters. It can be seen that by adding an SP Pack, sump tank sizes can be substantially reduced.

Step 14. Recovered oil tank. Assume a cylindrical tank with a retention time of 15 minutes and a process flow of 10% of the design water flow for flotation cells and a process flow of 5% of the design meter flow for skim vessels.

Field units

$$Q_w = (0.10)(5,000) + (0.05)(5,000)$$

$$= 750 \, BPD$$

$$= \frac{0.7(t_r)Q_w}{d_2},$$

$$H = \frac{7,875}{d_2},$$

$$H = \frac{0.7(15)(750)}{d^2}.$$

Vessel Diameter (in.)	Effective Length (ft.)	Seam-to-Seam Length (ft.)
30	8.8	11.8
36	6.1	9.1
42	4.5	7.5

SI units

$$Q_w = (0.10)(33) + (0.05)(33)$$
$$= 4.95 \, m^2/hr,$$
$$H = \frac{21,218(t_r)Q_w}{d^2},$$
$$H = \frac{21,218(15)(4.95)}{d^2}$$
$$= \frac{1,575,437}{d^2}.$$

Assume various diameters (d) and solve for liquid heights (H), $L_{ss} = L_{eff} + 3$ ft. $L_{eff} + 0.9$ m.

Vessel Diameter (in.)	Effective Length (ft.)	Seam-to-Seam Length (ft.)
762	2.7	3.6
914	1.9	2.8
1067	1.4	2.3

A vertical vessel 36 in. (914 mm) × 6 ft. (1.8 m) would satisfy all design parameters.

NOMENCLATURE

A_D = plan area of the deck, ft.^{2} (m^2)
C_i = inlet oil concentration
d = Vessel's internal diameter, in. (mm)
d = final droplet size, microns (μ)
d_b = diameter of gas bubble
d_m = oil droplet's diameter, microns (μ)
d_{max} = diameter of droplet above whose size only 5% of the oil volume is contained, microns (μ)
d_o = initial droplet size, microns (μ)
d_r = oil droplet diameter to be removed, microns (μ)
E = efficiency per cell
E_t = overall efficiency
F = factor that accounts for turbulence and short-circuiting
H = height of water, ft. (m)

h = height of mixing zone, ft. (m)

H_A = alarm level, ft. (m)

H_o = height of oil pad, ft. (m)

H_s = design annual storm surge, ft. (m)

H_{SD} = shutdown level, ft. (m)

H_T = normal tide range, ft. (m)

H_w = maximum height of oil below MWL, ft. (m)

j = an empirical parameter that is always larger than 3 and depends on the probability that the droplets will bounce apart before coalescence takes place

K_p = mass transfer coefficient

K_s = empirical parameter for the particular system

L = length of plate section parallel to the axis of water flow, ft. (m)

L = depth of pile below mean water level, ft. (m)

L' = length of baffle section, ft. (m)

L_{eff} = effective length in which separation occurs, ft. (m)

L_{ss} = seam-to-seam length, ft. (m)

N = number of plate packs

N = number of 50-gpm wash-down hoses

n = number of stages or cells

q_g = gas flow rate

Q_w = water flow rate, bwpd (m^3/hr)

q_w = liquid flow rate through mixing zone

Q_{wd} = wash-down rate, bwpd (m^3/hr)

r = radius of mixing zone

R_w = rainfall rate, in./hr (mm/hr)

SG_o = specific gravity of the oil relative to water

SG_w = specific gravity of the produced water

SG_w' = specific gravity of the sea water

$(t_r)_w$ = retention time, min

t_r = retention time, min

V_o = vertical velocity of the oil droplet relative to the water continuous phase, ft./s (m/s)

W = width, ft. (m)

α_w = fractional cross-sectional area of water

β_w = fractional water height within vessel

γ = height-to-width ratio, H/W

ΔP = pressure drip, psi (kPa)

ε = mixing parameter equivalent to the work done on a fluid per unit mass per unit time, cm^2/s^3

θ = angle of the plate with the horizontal

μ_w = water viscosity, cp (Pas)

ρ_w = water density, g/cm^3

σ = surface tension, dynes/cm

ϕ = volume fraction of the oil phase

Part 2
Water Injection Systems

Contents

INTRODUCTION

General Considerations

Oilfield waters usually contain impurities.

Impurities include:

Dissolved minerals

Dissolved gases

Suspended solids

Suspended solids can be:

Naturally occurring

Generated by precipitation of dissolved solids

Generated as products of corrosion

Created by microbiological activity

Precipitation of dissolved solids (scaling) is caused by changes in:

Temperature

Pressure

DOI: 10.1016/B978-1-85617-984-3.00002-X

pH

Or by mixing of waters from different sources

Suspended solids may:

Settle out of the water stream

Be carried as a suspension in flowing water

Primary sources of freshwater are:

Surface water

Ponds

Lakes

Rivers

Groundwater

Source water

In production operations, may come from

Separated produced water

Wells drilled into a subsurface water aquifer

May contain a large quantity of dissolved solids

Decision to use filtration should be a joint responsibility of the:

Reservoir engineer

Facilities engineer

This section provides information about equipment selection and sizing for removing suspended solids from water.

The water's source affects the types and amounts of contaminants in the water.

For example, produced water will be contaminated by some hydrocarbons.

The treatment of water to remove calcium and magnesium dissolved solids ("water softening") may be very important, especially if the water is to be used as boiler feed water for the generation of steam, as in a steamflood.

Nevertheless, a discussion of processes and equipment for water softening and removing other dissolved solids is beyond the scope of this section.

The removal of suspended solids from water may be desirable for a variety of reasons, the most common of which are to prepare the water for injection into a producing formation and to minimize the corrosion and solids build-up in surface equipment.

Prior to injecting water, it may be important to remove solids above a certain size to minimize damage to the formation caused by solids plugging.

This plugging can limit injection volumes, increase pump horsepower requirements, or lead to fracturing of the reservoir rock.

Dissolved gases such as oxygen in the water may promote bacteria growth within the formation, or they may speed the process of corrosion.

The presence of oxygen or hydrogen sulfide (H_2S) in water can lead to the formation of FeS, Fe_2, O_3 elemental sulfur particles, and scale.

These solid particles may form after the water is already downstream of solid removal equipment.

Without proper consideration of dissolved gases, the benefits of installing solids removal equipment can thus be partially negated.

In any solids removal system, there is a need for equipment to handle the bulk solids or sludge removed from the water and a procedure for removal of these solids.

For many common water injection systems, the amount of bulk solids to be removed can be large.

> If the solids are free of oil, they may be disposed of in a slurry piped to pits onshore or overboard offshore.

> If coated with oil, they may require treating prior to disposal.

Treating oil from solids is beyond the scope of this section.

Oil is normally separated from produced solids by abrasion in hydrocyclones or by washing with detergents or solvents.

Selection of a specific design of water-treating system for removing suspended gases from a water source requires establishing the year-round quality of the water source.

This determination normally requires that tests be performed to identify the amount of dissolved gases (primary), oxygen, and hydrogen sulfide (H_2S) present in the water, the total mass of suspended solids and their particle size distribution, and the amount of oil present in the source.

In addition, if a source of water is to be injected into a reservoir, it must be checked to ensure that it is compatible with reservoir water; that is, that under reservoir conditions, dissolved solids will not precipitate in sufficient quality to plug the well or reservoir.

Similarly, if two sources of water are to be mixed on the surface, they mus be checked for compatibility under surface conditions of pressure, temperature, and pH.

The tests are normally performed by laboratories that specialize in offering these services. (Determination of allowable concentration and particle size of solids and the acceptable level of dissolved gases in injection water is beyond the scope of this section.)

First, the theory involved in the various processes for removal of solids from water is discussed. Next, the equipment used is discussed, and finally, a design procedure to follow in selecting the equipment for a specific application is presented.

Treating water for solids removal and for removal of dissolved gases are really two separate concepts, using two separate sets of theories and equipment.

> They are usually considered together when designers plan a water-treating system for water injection.

> This section only discusses the removal of solids from water and not removing the dissolved gases.

Water softening, potable water making, and boiler feed water preparation are some of several important water treating topics that are not within the scope of this section.

Solids Content

For a variety of reasons, it may be desirable to remove suspended solids from a water stream.

> Removal is most commonly done as part of a water injection system for waterflood or enhanced oil recovery.

> It may also be necessary to remove suspended solids prior to injecting produced water in disposal wells.

Two different principles have been used to develop equipment for removing solids from water.

Gravity settling uses the density difference between the solid particle and the water to remove the solids

Filtration traps the solids within a filter medium that allows water to pass.

If the solids content of the water is low enough or if the solids are small in size, it may be possible to inject the water without:

Filtration

Excessive injection pressure

The number and size of the particles in the water along with the injectivity information of the formation:

Establishes whether filtration will be required

Dictates the type of filtration required

The quantity of suspended solids in a water stream is normally expressed in milligrams per liter (mg/l) or parts per million (ppm) by weight (mg/l times water specific gravity equals ppm).

The size of the suspended particles is usually expressed as a diameter stated in units of microns or micrometers (10^{-6} meters).

Oil Content

Removing the oil always aids the injectivity of the well.

Produced Water

Produced water usually contains enough solids to impair the permeability of sandstone formation whose permeability is less than 200 millidarcies.

The same water might be accepted in:

Vugular (voids or cavities) formation

Highly fractured limestones

Most produced waters require filtration to permit sustained injection.

Source Waters from Deep Sand Formation

These usually need filtration for at least a few weeks after completion of the well to determine if the water will become free of sediments.

If the water is determined to be sediment-free:

Filtration can probably be stopped.

Close surveillance should be maintained until it is thoroughly established that filtration is no longer needed.

Filtration should be resumed temporarily each time:

Source well pump is pulled

Other work is performed in the well

There are several types of filters and sediment removal devices available.

SOLIDS REMOVAL THEORY

Removal of Suspended Solids from Water

Removal of suspended solids is commonly required in the following scenarios:

Water injection system for water-flood

Enhanced oil recovery

Prior to injecting produced water in disposal wells

Suspended solids are removed by:

Filtration

Gravity settling

Uses the density difference between the solid particles and water to remove the solids.

Filtration

Traps solid particles of a specified size within a filter media.

Allows passage of water.

Quantity of suspended solids in a water stream is normally expressed in:

Milligrams per liter (mg/l)

Parts per million (ppm) by weight

ppm = mg/l divided by SG_{water}

Size of the suspended particles are expressed as a diameter stated in:

Micrometers (10^{-6} meters)

Also called microns

Capacity of equipment or filters to remove suspended solids is expressed in terms of:

Removal of a percentage of all suspended solids having a diameter greater than a specified micron size

Values will usually range from:

150 microns (μ) for gravity separators

Less than 0.5 microns (μ) for filters

Suspended solids less than 40 microns (μ) in diameter cannot be seen with the naked eye.

Figure 2-1 shows relative sizes for a variety of common materials.

Gravity Settling

Solids have a density greater than water, therefore they fall relative to the water due to the force of gravity.

The terminal settling velocity is such that the gravitational force on the particle equals the drag force resisting its motion due to friction.

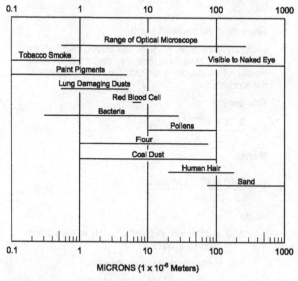

FIGURE 2-1 Relative sizes of common materials.

Assuming the particle is roughly spherical, the drag force may be determined as follows:

$$F_D = C_D A \left| \frac{V_t^2}{2_g} \right| \tag{2-1}$$

where

F_D = drag force, lb (kg)

C_D = drag coefficient

A = cross-sectional area of the particle, ft.^{2} (m^2)

ρ = density of the continuous phase, lb/ft.^{3} (kg/m^3)

V_t = terminal settling velocity of the particle, ft./s (m/s)

G = gravitational constant, 32.2 ft./s^2 (9.81 m/s^2)

The terminal settling velocity of small particles through water is low, and flow around the particle is laminar.

Therefore, Stokes' law may be applied to determine the drag coefficient as follows:

$$C_D = \frac{24}{Re} \tag{2-2}$$

where Re is the Reynolds number.

By equating the gravitationally induced negative buoyant force with the drag force, one may derive the following equation for calculating the terminal settling velocity of the particle

Field units

$$V_t = \frac{1.78 \times 10^{-6} (\Delta SG) d_m^2}{\mu} \tag{2-3a}$$

SI units

$$V_t = \frac{5.44 \times 10^{-10} (\Delta SG) d_m^2}{\mu} \tag{2-3b}$$

where

ΔSG = difference in specific gravity of the particle and the water

d_m = particle diameter, microns (μ)

μ = viscosity of the water, cp (Pas)

Equation 2-3a or 2.3b may be used to size any of several types of equipment designed to use gravity settling. Equipment includes:

Onshore (where space is available)

> Settling ponds
>
> Pits
>
> Flumes
>
> Tanks

Offshore

> PPIs
>
> CPIs
>
> Cross-flow separators
>
> Hydrocyclones
>
> Centrifuges
>
> Flotation units

Gravity settling devices:

> Remove large particles (greater than 10 microns)
>
> Require low velocities and long retention times
>
> Are effected by chemical additions
>
> Have system inefficiencies that cause oversizing

Filtration

Filtration is used to remove suspended solid particles from water by passing the water through a porous filter medium.

Particles larger than the pores become trapped.

Size of the pores in a filter medium determines the smallest particles that may be trapped.

Suspended solids are separated from fluids via three mechanisms:

> Inertial impaction
>
> Differential interception
>
> Direct interception

Inertial Impaction (Figure 2-2)

Particles (1 to 10 microns) in a fluid stream have mass and velocity and, hence, have a momentum associated with them.

As the liquid and entrained particles pass through a filter media:

Liquid stream will take the path of least resistance to flow and will be diverted around the fiber.

Particles, because of their momentum, tend to travel in a straight line and, as a result, those particles located at or near the center of the flow line will strike or impact upon the fiber and be removed.

FLUID FLOW STREAMLINES

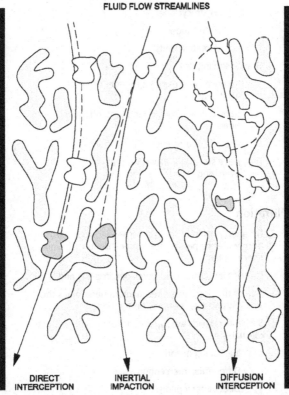

| DIRECT INTERCEPTION | INERTIAL IMPACTION | DIFFUSION INTERCEPTION |

FIGURE 2-2 Filtration mechanisms.

This is not effective in liquid filtration since the differential densities of the particles and fluids are very small and thus the deviation from the liquid flow line is much less.

Diffusional Interception

In particles that are extremely small (i.e., those with very little mass and less than 0.3 microns in diameter), separation can result from diffusional interception.

Particles are in collision with the liquid molecules.

These frequent collisions cause the suspended particles to move in a random fashion around the fluid flow lines.

Such motion is called *Brownian motion*, which causes these smaller particles to deviate from the fluid flow lines and hence increase the likelihood of their striking the fiber surface and being removed.

This is not effective in liquid filtration because of the inherent nature of liquid flow, which tends to reduce the lateral movement of the particle away from the fluid flow lines.

Direct Interception

Direct interception is equally effective in both liquid and gas service.

It is the desired mechanism for separating particles (0.3 to 1 micron range) from liquids.

The filter medium has an assembly of a large number of fibers which define openings (pores) through which the fluid passes.

Particles larger than the openings (pores) will be removed.

Filters collect a significant proportion of particles whose diameter is smaller than the openings (pores) on the medium.

Factors that account for this collection are:

Most suspended solids are irregular in shape and hence can "bridge" an opening.

Bridging effect also occurs if two or more particles strike an opening simultaneously.

Once a particle has been stopped by a pore, that pore is at least partially occluded and subsequently will be able to separate even smaller particles from the liquid stream.

Specific surface interactions can cause a small particle to adhere to the surface of the internal pores of the medium.

A particle considerably smaller than a pore is likely to adhere to that pore provided the two surfaces are oppositely charged.

A strong negatively charged filter can cause a positive charge to be induced on a less strongly charged negative particle.

FILTER TYPES

Nonfixed-Pore Structure Media

These depend principally on the filtration mechanisms of inertial impaction and/or diffusional interception.

They are constructed of nonrigid media.

Variations in pressure drop may cause minor deformation or movement of the filter medium, potentially changing the size of some of the pores.

These depend not only on trapping but also on adsorption to retain particles.

When a filter has been on stream for a length of time and has collected a certain amount of particulate matter, a sudden increase in flow and/or pressure can cause the release (unloading) of some of the particles.

These are the most common type of filter and include the following:

Unbonded fiberglass cartridges

Cotton wound or sock filters

Molded cellulose cartridges

Spun wound polypropylene cartridges

Sand and other granular media beds

Diametomaceous earth filters

Fixed-Pore Structure Media

These consist of either layers of medium or a single layer of medium having depth.

They depend heavily on the mechanism of direct interception.

They are constructed such that:

> Structural portions of the medium cannot distort and thus the flow path through the medium is tortuous.

> Pore size does not change.

They yield more consistent particle removal efficiencies.

They experience lower solids unloading.

This represents new technology in filter medium construction for oilfield use and include the following:

> Resin impregnated cellulose cartridges

> Resin-bonded glass fiber cartridges

> Continuous polypropylene cartridges

Eventually a filter collects solids until the pressure drop is too large for continued operation.

> At this point, the filter must be replaced or cleaned.

> Amount of solids a filter removes per volume is referred to as a filter's "solids loading."

Different types of filters have different solids loading capabilities.

> Filter's solids loading capability is affected by the filter design and particular medium used.

Figure 2-3 shows sections of three filters.

> Using different fiber media, which can be glass fiber, cellulose, cotton, or polypropylene.

> All have the same filter area with the same pore size.

> Only difference is the fiber diameter used to form the medium.

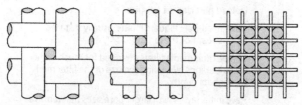

FIGURE 2-3 Fiber diameter affects filter's solids capacity.

Right side represents a filter with 16 times as many pores per volume as the filter on the left, thus its solids loading should be much larger.

Surface Media

Surface, or screen media, is one in which all pores rest on a single plane, which therefore depends largely upon direct interception. Examples include:

Woven wire mesh

Woven cloth

Membrane filter

It stops all particles larger than the largest pore opening.

Particles smaller than the largest pore may be stopped because of bridging, and suchlike. Thus, there is no guarantee that such particles will not pass downstream.

Woven wire mesh filters are currently available with openings down to 5 microns.

Summary of Filter Types

A meaningful classification of filters is as follows:

1. Nonfixed-pore structures have pores whose dimensions increase at high pressure ("wound," low-density filters).

2. Fixed-pore structures have pores that do not increase in size at high pressures (most membrane filters).

3. Screen media (woven cloth or screens).

Nonfixed-pore structure filters:

Do not have absolute ratings

Are subject to media migration

Unload particles on impulse

Have fixed-pore "guard" filter usually installed downstream of these filters

Can be multimedia filters designed for much higher solids loadings by varying the size of the filter media through its depth

Have initial layers with larger pore sizes that remove the greatest weight of solids (those having the largest diameters) while successive layers remove smaller and smaller diameter solids from the flow stream

Fixed-pore structure filters:

Are superior when compared to screen filters

Combine high dirt capacity per unit area with both absolute removal of particles larger than a given size and minimum release of collected particles smaller than this rated size under impulse conditions

REMOVAL RATINGS

General Considerations

Attention should be paid to filter rating.

No generally accepted rating system.

Several of the rating systems are described below.

Nominal Rating

The National Fluid Power Association (NFPA) states

An arbitrary micron value assigned by the manufacturer, based upon removal of some percentage of all particles of a given size or larger. It is rarely well defined and not reproducible.

A "contaminant" is introduced upstream of the filter element

Subsequently the fluid flow (flow downstream of the filter) is analyzed microscopically

A given nominal rating of a filter means 98% by weight of the contaminant above the specified size has been removed; 2% by weight of the contaminant has passed downstream.

This is a gravimetric test rather than a particle count test.

Counting particles upstream and downstream is a more meaningful way to measure filter effectiveness.

Tests used to give nonfixed-pore structure filters a nominal rating yield results that are misleading. Typical problems are as follows:

1. 98% contaminant removal by weight is determined by using a specific contaminant at a given concentration and flow. If any one of the test conditions is changed, the test results could be altered significantly.

2. 2% of the contaminant passing through the filter is not defined by the test. It is not uncommon for a filter with a nominal rating of 10 μ to pass particles downstream ranging from 30 to over 100 μ.

3. Test data are often not reproducible, particularly among different laboratories.

4. Some manufacturers do not base their nominal rating on 98% contaminant removal by weight, but instead a contamination removal efficiency of 95%, 90%, or even lower. Thus, it often happens that a filter with an absolute rating of 10 μ is actually finer than another filter with a nominal rating of 5 μ. Therefore, it is always advisable to check the criteria upon which a nominal rated filter is based.

5. The very high upstream contaminant concentrations used for such tests are not typical of normal system conditions and produce misleading high-efficiency values. It is common for wire-mesh filter medium with a mean (average) pore size of 5 μ to pass a 10 μ nominal specification. However, at normal system contaminants, the same filter medium will pass almost 10 μ particles.

Therefore, one cannot assume that a filter with a nominal rating of 10 μ will retain all or most of the particles 10 μ or larger.

Yet some filter manufacturers continue to use only a nominal rating because it:

Makes their filters seem finer than they actually are

It's impossible to place an absolute rating on a nonfixed-pore structure

Absolute Rating

NFPA defines an absolute rating as follows:

The diameter of the largest hard spherical particle that will pass through a filter under specified test conditions. It is an indication of the largest opening in the filter element.

Such a rating can be assigned only to an integrally bonded medium.

There are a number of recognized tests used.

Which test is used depends on the:

> Manufacturer
>
> Type of medium to be tested
>
> Processing industry

Filters are rated by a "challenge" destructive test.

> Filter is challenged by pumping through a suspension of a readily recognized contaminant and both the influent and effluent examined for the presence of the test contaminant.
>
> Filter cannot be used after the test.

Nondestructive integrity tests have been established:

> Correlate with destructive challenge test
>
> Filter element can be placed in service after the test

Beta (β) Rating System

Determined using the Oklahoma State University *OSU F-2 Filter Performance test*.

Used to measure and predict the performance of a wide variety of filter cartridges under specified conditions.

Based on measuring the total particle counts at several different particle sizes, in both the influent and effluent streams.

A profile of removal efficiency is then given for the filter.

The Beta ratio is defined as follows:

$$B_x = \frac{\text{\# of particles} \geq X \text{ in influent}}{\text{\# of particles} \geq X \text{ in effluent}}$$

where X is the particle size, in microns

The percent removal efficiency at a given particle size can be obtained directly from the Beta ratio and can be calculated as follows:

$$\%\text{Removal efficiency} = \left(\frac{B_x - 1}{B_x}\right) 100$$

where:

> B_x = beta ratio
>
> X = particle size, microns

Table 2-1 Beta Ratio and Removal Efficiency Comparison

No. of Particles per ml $\geq$ 10 m			Beta Ratio B10	Removal Efficiency %
Filter	Influent	Effluent		
A	10,000	5000	2	50
B	10,000	100	100	99
C	10,000	10	1000	99.9
D	10,000	2	5000	99.98
E	10,000	1	10,000	99.99

The relationship between Beta ratio and percent removal efficiency is illustrated in Table 2-1.

Usually $B_x = 5000$ can be used as an operational definition of absolute rating.

The Beta values allow comparisons of removal efficiencies at different particle sizes for different cartridges in a meaningful manner.

The following parameters are important when selecting a filter:

Type of filter medium

Rating (nominal, absolute, or Beta)

Solids loading

The designer should pay attention to the precise meaning of the manufacturer's rating.

CHOOSING THE PROPER FILTER

General Considerations

Important factors that must be taken into consideration when choosing a filter for a particular application are:

Size, shape, and hardness of the particles to be removed

Quantity of the particles

Nature and volume of the fluid to be filtered

Rate at which the fluid flows

Whether the flow is steady, variable, and/or intermittent

System pressure and whether that pressure is steady or variable

Available differential pressure

Compatibility of the medium with the fluid

Fluid temperature

Properties of the fluid

Space available for particle collection

Degree of filtration required

Nature of Fluid

Materials from which the medium, the cartridge hardware, and the housing are constructed must be compatible with the fluid being filtered.

Fluids can corrode the metal core of a filter cartridge or a pressure vessel and corrosion will in turn contaminate the fluid being filtered.

Flow Rate

Flow rate through the filter is dependent upon

Pressure drop available (ΔP)

Resistance to flow at the filter media (R)

If the pressure drop available is increased then the flow rate of that fluid through the media will increase.

As viscosity increases, the pressure required to maintain the same flow rate increases.

Temperature

Temperature at which filtration occurs can effect:

Viscosity of the fluid

Corrosion rate of the housing

Filter medium compatibility

It's important to determine the viscosity of a fluid at the temperature at which filtration will occur.

High temperature:

Accelerates corrosion

Weakens the gaskets, and seals of the filter housing

Disposable filter media cannot withstand high temperature for long periods of time.

Often one must choose porous metal, cleanable filters.

Pressure Drop

Differential pressure:

Moves the fluid through the filter assembly

Overcomes resistance to flow and ΔP

Filtration systems must provide a sufficient pressure source to:

Overcome the resistance of the filter

Permit flow to continue at an acceptable rate as the medium plugs so as to use fully the effective solids holding capacity of the filter

Maximum allowable pressure drop:

Has a limit beyond which the filter might fail structurally should additional system pressure be applied to maintain flow

Is specified by the filter manufacturer

Surface Area

The life of most screen and fixed-pore structure filters is greatly increased as their surface areas are increased.

An increase in surface area will yield at least a proportional increase in service life.

The ratio of service life may approach the square of the area ratio.

A filter user will save money in the long run by paying the higher initial cost of a larger filter assembly.

Void Volume

The medium with the greatest void volume is desirable because it:

Yields the longest life

Has the lowest initial clean pressure drop per unit thickness

As fiber diameter decreases, void volume increases, assuming constant pore size.

Other factors must be considered when designing a filter for a particular application:

Strength

Compressibility as pressure is applied (which reduces void volume)

Compressibility of the medium with the fluid being filtered

Cost of medium

Cost of constructing that medium into a usable filter

Degree of Filtration

Filter must be able to remove contamination from the fluid stream to the degree required by the process involved.

Once the size of the contaminants to be removed has been determined:

It is possible to choose a filter with the particle removal characteristics needed to do the job.

Choosing a filter with a pore size finer than required can be a costly mistake.

The finer the filtration the:

More rapid the clogging

Higher the cost

Filter selected must be able to retain particles removed from the subject fluid.

Depth-type filters of the type whose pores can increase in size as pressure is increased are subject to unloading.

With surface filters or fixed structure filters, one selects a medium that will not change its structure under system-produced stress.

Prefiltration

Prefiltration removes bulk quantities of significantly higher size particles.

It reduces overall operating cost by extending the life of the final filter.

It extends final filter life.

May not in itself be sufficient to justify prefiltration

Overall cost reduction is usually the principle consideration

Field experience indicates that for most applications where the solids are of near uniform size it is better to increase the final filter area rather than provide a prefilter.

Increasing the final filter area always yields:

Longer cycle time

Lower operating costs

Doubling the area of the final filter will result in two to four times the life.

On the other hand:

Installing settling devices, hydrocyclone desanders, or large pore space sand filters upstream of filters designed to remove very small particles is often a more economical solution than just increasing the size of the final "polishing" filter.

Coagulants and Flocculation

Suspended matter in water may contain very small particles that:

Will not settle out by gravity

May pass through filters

Particles may be removed by a coagulation-and-flocculation process.

Coagulation is the process of destabilization by charge neutralization.

Flocculation is the process of bringing together the destabilized or coagulated particles to form a larger agglomerate, or floc.

Coagulation and flocculation results are:

Difficult to predict based on a water analysis

Laboratory jar tests are:

Performed to simulate the coagulation and flocculation condition

Data used to determine the basis for design and efficient operation

Tests that establish

Optimum pH for coagulation

Most effective coagulation and coagulation aid

Most effective coagulation dosage and order of chemical addition

Coagulation and flocculation time

Settling time or flocculation time

Chemicals used include:

Chlorine

Bentonite (for low-turbidity water)

Primary inorganic coagulants

pH adjusted chemicals

Polyelectrolytes

Chlorine whose addition may assist coagulation by oxidizing organic contaminants that have dispersing properties

Waters with high organic content, which require high coagulant demand

Chlorination prior to addition of coagulant feed that may reduce the required coagulant dosage

Polyelectrolytes are characterized by the following:

Refer to all water-soluble organic polymers used for clarification of water by coagulation

Available water-soluble polymers may be classified as

Anionic

Cationic

Approaching neutral charge

Are typically long-chain, high-molecular-weight polymers with many charge sites to aid in coagulation and flocculation

Violent mixing of polyelectrolytes may break the chains and cause them to be less effective

Some mixing is required to ensure that the chemical and solids come into contact

Turbulent flow in piping provides sufficient mixing if the chemical is injected far enough upstream of the equipment

Chemical addition require:

Water clarifiers (tanks with mixers that cause turbulence to create contact between the chemical and solids)

Flotation units to aid in attaching gas bubbles to the solid particles

Feed stream to filtration units that increase filtration efficiency

MEASURING WATER COMPATIBILITY

Scale deposits are usually salts or oxides of calcium, magnesium, iron, copper, and aluminum.

Common scales deposits may consist of:

Calcium carbonate ($CaCO_2$)

Calcium phosphate ($CaPO_4$)

Calcium silicate ($Ca_2Si_2O_4$)

Calcium sulfate ($CaSO_4$)

Magnesium hydroxide [$Mg (OH)_2$]

Magnesium phosphate ($MgPO_4$)

Magnesium silicate ($Mg_2Si_2O_4$)

The tendency of water to form scale or cause corrosion is measured by either:

Langelier Scaling Index (LSI)

Also called the Saturation Index (Table 2-2)

Ryznar Stability Index (RSI)

Also called the Stability Index

Saturation Index (LSI)

Deals with the conditions at which given water is in equilibrium with calcium carbonate and provides a qualitative indication of the tendency of calcium carbonate to deposit or dissolve.

Index is determined by subtracting from actual pH of the water sample (pHA), a computed value (pHS) based on the ppm of calcium hardness as $CaCO_3$, alkalinity hardness as $CaCO_3$, and total solids, as shown in Figure 2-4. Refer to the example in Table 2-2.

If the index is positive, calcium carbonate will tend to deposit.

If the index is negative, calcium carbonate will tend to dissolve.

Table 2-2 Saturation Index

To determine:	
pCa	Locate ppm value for Ca as $CaCO_3$ on the ppm scale. Proceed horizontally to the left diagonal line down to the pCa scale.
pAlk	Locate ppm value for "M" Alk as $CaCO_3$ on the ppm scale. Proceed horizontally to the right diagonal line down to the pAlk scale.
Total solids	Locate ppm value for total solids on the ppm scale. Proceed horizontally to the proper temperature line and up to the "C" scale.

Example:
Temp. = 140 °F, pH = 7.80   pCa = 2.70
Ca hardness = 200 ppmw   pAlk = 2.50
M alkalinity = 160 ppmw   C at 140°F = 1.56
Total solids = 400 ppmw   Sum = pH 3 = 6.76
                              Actual pH = 7.50
                              Difference = 1.04

Stability Index (RSI)

Given by the following equation

$$RSI = 2Ph_s - pH_A$$

When the index is

Less than 6, scaling can be expected

Between 6 and 7 indicates a stable water

Greater than 7 indicates potential corrosion problems

Scaling may be controlled by the following:

Blow-down to limit build-up of solids concentrations

Acid treatment to reduce the water alkalinity

Use of commercial scale inhibitors such as polyelectrolytes, phosphonates, and polymers

SOLIDS REMOVAL EQUIPMENT DESCRIPTION

Source Water Considerations

Water Injection Treatment Steps

Figure 2-5 shows a schematic of steps that may be required to prepare water for injection.

Choice of process is affected by the water source.

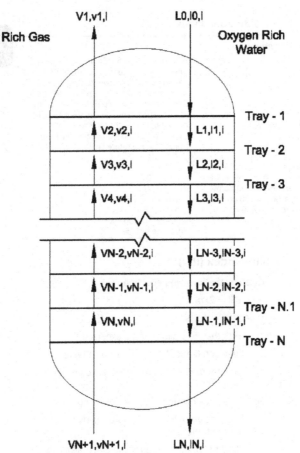

FIGURE 2-4 Langelier Saturation Index Chart. Reprinted from *GPSA Engineering Data Book*, courtesy of Betz Laboratories, Inc.

Produced Water

May be disposed of once its hydrocarbon content is reduced to acceptable levels (25 to 50 mg/l).

When water is to be injected:

It may require filtration to remove dispersed oil.

It prevents impairment of the injection formation.

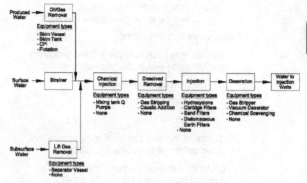

FIGURE 2-5 Water injection system treatment steps and equipment types.

Water must be cleaned to less than 50 mg/l suspended oil so that oil does not create plugging problems in the filters.

Surface Water

Common source for water-floods and other injection projects.

May be cheaper than subsurface water.

May require more treatment than other water sources

Since it is fresh, it may cause swelling of clays in some formations.

It should be free of large contaminants such as plant or marine life.

> Strainer in Figure 2-5 is intended to prevent such material from entering treatment facilities.

Surface water is exposed to the atmosphere and thus contains dissolved oxygen.

> Oxygen should be removed to minimize corrosion and bacteria growth.

> Oxygen concentration within the water will vary, depending on the water temperature.

> De-aeration equipment must be designed to remove the maximum anticipated oxygen.

Oxygen concentrations of approximately 8 ppm are typical for most surface waters.

Chemical injection is required.

Biocides may be used to prevent the growth of the following within the treating system:

Microscopic marine life

Bacteria

Plankton

Corrosion inhibitors may be used to minimize deterioration of the surface facilities.

Bactericide may be used on either a periodic or a continuous basis to minimize growth.

Chemicals such as oxygen scavengers may be injected to react with contaminants to remove or change them.

Chemicals may be injected to prevent the:

Formation of scale within the surface equipment

Precipitation of solids under reservoir conditions

Subsurface Water

Source water wells may also be used.

Water zone will be close to the surface, compared with hydrocarbon zones.

Cost of drilling and maintaining source water wells is typically much lower than the same cost for producing wells.

Water will rarely flow to the surface under pressure, thus it must be pumped or gas lifted.

If water is gas-lifted to the surface, separation facilities will be required.

Two-phase separator is sufficient since small levels of dissolved gas in the water are not harmful to the surface equipment or the injection formation.

If gas lift gas contains acid gases or oxygen, treating may be necessary to remove or neutralize contaminants.

Subsurface water is the least expensive source water to treat.

However, the cost of drilling source water wells and the pumping or gas lifting expenses may make this the most expensive water to obtain.

Dissolved Minerals and Salts

Minerals and salts are contained in all source waters.

They will remain in the solution and thus are not a problem.

Water Compatibility

If produced water is mixed with other source water as part of a water flood, the water compatibility should be verified.

Mixing other waters with produced water or changing the produced water's pH or temperature may cause dissolved solids to precipitate.

Compatibility of the injected water and the water in the injection formation should also be checked to ensure that the conditions within the reservoir will not cause scale to form.

Chemicals may be needed to inhibit scale formation under down-hole conditions.

Filtration

Removes suspended solids, no matter what the source, and is intended to minimize plugging of the formation.

Solids that are in the injected water and are larger than a certain size may plug the formation at the well bore, causing the surface injection pressure to rise or the flow rate injected to fall.

Degree of filtration required depends on the permeability and pore size of the injection formation.

Final selected filtration design should be one that will:

Minimize formation plugging

Reduce the frequency of remedial well work

Selection is an economic decision balancing the cost of remedial well work against the cost of the fiber system.

When water is injected in a disposal well, filtration requirements might be relaxed.

Disposal wells are typically less expensive to drill or work over and more readily fractured than wells injecting the water into a producing formation.

Economic risks of plugging a disposal well may be less than those associated with a well injecting the water into the producing formation.

The test to check for compatibility of injection water with the receiving formation may consist of:

A chemical analysis of the proposal injection water is required to indicate basic cations and anions present.

A core plug test used to determine the maximum particle size that can be injected into the formation without undue plugging.

A core plug test to determine and establish permissible injection rates and pressures.

Gravity Settling Tanks

Simplest treating equipment used to remove solids from water.

Configurations include:

Vertical (Figure 2-6)

Horizontal (Figure 2-7)

Vertical Vessels

Water enters the vessel and flows upward to the water outlet.

Solids fall countercurrent to the water and collect in the bottom.

Large diameter vessels or tanks should have spreaders and collectors to distribute the water flow and minimize short-circuiting.

Water outlet is high thus allowing better removal of solids from the liquid.

Cone bottoms are used rather than elliptical heads as they allow more complete removal of solids through the drain.

An angle of 45 to 60° from the horizontal is used for the cone to overcome the angle to overcome the

2

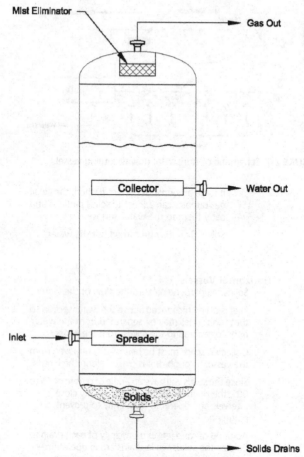

FIGURE 2-6 Schematic of vertical gravity settling vessel.

resistance of the sand and allow a natural flow of solids when the drain is opened.

Any flash gas that evolves from the water leaves the settling vessel through the gas outlet at the top of the vessel.

Volume of flash gas must be kept to a minimum so the gas does not adversely affect the removal of small solids particles.

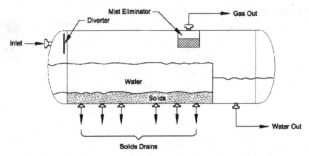

FIGURE 2-7 Schematic of horizontal gravity settling vessel.

If large amounts of gas are flashed, the small gas bubbles can adhere to solids particles and carry them to the water surface.

Solids then may be carried out the water outlet.

Horizontal Vessels

Solids fall perpendicular to the flow of the water.

Inlet is often introduced above the water section so that flash gases may be separated from the water prior to separating the solids from the water.

Collected solids must be periodically removed from the vessel through drains placed along the length.

Since the solids will have an angle of repose of 45 to 60°, the drains must be spaced at very close intervals and operated frequently to prevent plugging.

Addition of sand jets in the vicinity of each drain to fluidize the solids while drains are in operation is expensive, but sand jets proved successful in keeping drains open.

Alternatively, the vessel may have to be shut down so that solids may be manually removed through a manway.

This is more efficient at solids separation since the solid particles do not have to fall countercurrent to the water flow.

Other considerations, such as the difficulty of removing solids, must be considered when selecting a configuration.

Horizontal vessel considerations:

> Better from a process point of view
>
> Require more plan area to perform the same separation as a vertical
>
> Small vessels have less liquid surge capacity, which require LSHs to be set closer to the normal operating level

Vertical vessel considerations:

> Controls and PSV may be difficult to service without special ladders and platforms.
>
> Vessels must be removed from skid when shipping.

Pressure vessels are more expensive than tanks.

Pressure vessels should be considered when:

> Potential blow-by through an upstream vessel dump system could create too much backpressure in an atmosphere tank's vent system, or
>
> Water must be dumped to a higher elevation for further treating and a pump would be needed if an atmospheric tank were installed

For gravity settling of solids:

> Retention time does not directly affect the solids removal
>
> 30 seconds are required for the evolved gases to flash out of solution and reach equilibrium

Only settling theory must be considered. Gravity settling theory is used where there is a high solids flow rate of large-diameter (greater than 50-microns) solids that might otherwise quickly overload equipment designed to separate smaller-diameter solids from the liquid stream.

Horizontal Cylindrical Gravity Settlers

The required diameter and length of a horizontal cylindrical settler can be determined from Stokes' law as follows:

Field units

$$dL_{eff} = 1000 \frac{\beta_w Q_w \mu_w}{\alpha_w (\Delta SG) d_m^2} \qquad (2\text{-}4a)$$

SI units

$$dL_{eff} = 1000 \frac{\beta_w Q_w \mu_w}{\alpha_w (\Delta SG) d_m^2} \qquad (2\text{-}4b)$$

where

d = vessels's internal diameter, ft. (m)

L_{eff} = effective length in which separation occurs, ft. (m)

Q_w = water flow rate, bwpd (m³/hr)

μ_w = water viscosity, cp

d_m = particle diameter, microns μ

ΔSG = difference in specific gravity between the particle and water relative to water

β_w = fractional water height within the vessel (h_w/d)

α_w = fractional cross-sectional area of water

h_w = water height, in. (m)

Equation 2-4 assumes a turbulence and short-circuiting factor of 1.8.

Any combination of d and L_{eff} that satisfies this equation will be sufficient to allow all particles of diameter d_m or larger to settle out of the water.

The fractional water height and fractional water cross-sectional areas are related by the following equation:

$$\alpha_w = (1/180) \cos^{-1}[1 - 2\beta_w] - (/\pi)[1 - 2\beta_w] \\ \sin[\cos^{-1}(1 - 2\beta_w] \qquad (2\text{-}5)$$

By selecting a fractional water height within the vessel, one may calculate the associated fractional cross-sectional area using Equation 2-5, the resulting values may then be used in Equation 2-4.

In addition to the settling criteria, a minimum retention time (less than 30 seconds) should be provided to allow the water and flash gases to reach equilibrium.

To ensure that the approximate retention time has been provided, the following equation must also be satisfied when selecting d and L_{eff}.

Field units

$$d^2 L_{eff} = \frac{(t_r)_w Q_w}{1.4\alpha_w} \quad\quad (2\text{-}6a)$$

SI units

$$d^2 L_{eff} = 21,000 \frac{(t_r)_w Q_w}{1.4\alpha_w} \quad\quad (2\text{-}6b)$$

The choice of the correct diameter and length can be obtained by selecting various values for d and L_{eff} for both Equations 2-5 and 2-6.

For each d the larger L_{eff} must be used to satisfy both equations.

The relationship between the L_{eff} and the seam-to-seam length of a settler is dependent on the settling vessel's internal physical design.

Some approximations of the seam-to-seam length may be made based on experience as follows:

$$L_{ss} = (4/3)L_{eff} \quad\quad (2\text{-}7)$$

where L_{ss} = seam-to-seam length.

This approximation must be limited in some cases, such as vessels with large diameters. Therefore, the L_{ss} should be calculated using Equation 2-7, but it must exceed the value calculated using the following equations:

Field units

$$L_{ss} = L_{eff} + 2.5 \quad\quad (2\text{-}8a)$$

where $L_{eff} < 7.5$ ft.

SI units

$$L_{ss} = L_{eff} + 0.76 \qu\quad (2\text{-}8b)$$

Field units

$$L_{ss} = L_{eff} + (d/24) \quad\quad (2\text{-}9a)$$

SI units

$$L_{ss} = L_{eff} + (d/2000) \quad\quad (2\text{-}9b)$$

Equation 2-8a and 2.8b only govern when the calculated L_{eff} is less than 7.5 ft. (2.3m).

The justification of this limit is that some minimum vessel length is always required for smoothing the water inlet and outlet flow.

Equation 2-9 governs when one-half the diameter in feet exceeds one-third of the calculated L_{eff}. This constant ensures that even flow distribution can be achieved in short vessels with large diameters.

Horizontal Rectangular Cross-Sectional Gravity Settlers

The required width and length of a horizontal tank of rectangular cross section can be determined from Stokes' law as:

Field units

$$WL_{eff} = 70 \frac{Q_w \mu_w}{(\Delta SG) d_m^2} \qquad (2\text{-}10a)$$

SI units

$$WL_{eff} = 9.7 \times 10^5 = \frac{Q_w \mu_w}{(\Delta SG) d_m^2} \qquad (2\text{-}10b)$$

where

W = width, ft. (m)

L_{eff} = effective light in which separation occurs, ft. (m)

Equation 2-10a and 2.10b:

Assume a turbulence and short-circuiting factor of 1.9

Is independent of height because the particle settling time and the water retention time are both proportional to the height

Typically, the height is limited to one-half of the width to promote even flow distribution.

An equation may be developed to ensure that sufficient retention time is provided. If the height-to-width ratio is set, then the following retention time equation applies:

Field units

$$W^2 L_{eff} = \frac{0.004(t_r)_w Q_w}{\gamma} \qquad (2\text{-}11a)$$

SI units

$$W^2 L_{eff} = \frac{(t_r)_w Q_w}{60\gamma} \qquad (2\text{-}11b)$$

where

$$\gamma = \text{height-to-width ratio, } (H_w/W)$$

$$H_w = \text{height of the water, ft. (m)}$$

As with horizontal cylindrical settlers, the relationship between L_{eff} and L_{ss} depends on the internal design.

Three approximations of the L_{ss} of rectangular settling vessels may be made using Equation 2-4. However, the L_{ss} must be limited by Equation 2-5 and the following:

$$L_{ss} = L_{eff} + W/2 \qquad (2\text{-}12)$$

As before, the L_{ss} should be the larger of Equations 2-7, 2-8, and 2-12.

Vertical Cylindrical Gravity Settlers

The required diameter of a vessel cylindrical tank can be determined by setting the settling velocity equal to the average water velocity as follows:

Field units

$$d^2 = 6700F \frac{Q_w \mu_w}{(\Delta SG) d_m^2} \qquad (2\text{-}13a)$$

SI units

$$d^2 = 6.5 \times 10^{11} F \frac{Q_w \mu_w}{(\Delta SG) d_m^2} \qquad (2\text{-}13b)$$

where

$F =$ factor that accounts for turbulence and short-circuiting $= 1.0$ (for diameters < 48 in. (1.22 m) $= d/48$ (for diameters > 48 in. (1.22 m)

Substituting $f = d/48$ into Equation 2-13 gives the following:

Field units

$$d = 140 \frac{Q_w \mu_w}{(\Delta SG) d_m^2} \qquad (2\text{-}14a)$$

SI units

$$d = 5.3 \times 10^9 \frac{Q_w \mu_w}{(\Delta SG) d_m^2} \qquad (2\text{-}14b)$$

where

$$d = 48 \text{ in. (1.22 m)}$$

Equation 2-14 applies only if the settler diameter is greater than 48 in. (1.22 m).

For smaller settlers, Equation 2-13 should be used and F should equal 10.

The height of the water column in feet can be determined for a selected d from retention time requirements:

Field units

$$H = 0.7 \frac{(t_r)_w Q_w}{d^2} \tag{2-15a}$$

SI units

$$H = 21,000 \frac{(t_r)_w Q_w}{d^2} \tag{2-15b}$$

where H is the height of water, ft. (m).

Plate Coalescers

Equations for sizing the various configurations are identical to those presented in Part 1 and can be used directly, where d_m is the diameter of the solid particle (and not the oil droplet diameter) and ΔSG is the difference in specific gravity between the solid and water (and not between oil and water).

Plate coalescers are not addressed in this section because they have a tendency to plug and thus are not recommended for solids.

Hydrocyclones

Also called desanders or desilters.

Direct water into a cone through a tangential inlet that imparts rotational movement to the water.

Figure 2-8a and 2-8b show a hydrocyclone cone assembly and an assembly of eight cones.

Rotary motion generates centrifugal forces toward the outside of the cone, which drives the heavy solids to the outer perimeter of the cone.

Once the particles are near the wall, gravity draws them downward to be rejected at the apex of the cone.

The resulting heavy slurry is then removed as "underflow."

Clear water near the center of the vertical motion is removed through an insert at the center of the

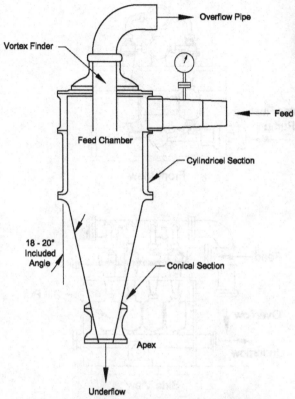

FIGURE 2-8A Schematic of a hydrocyclone core assembly.

centerline of the hydrocyclone, called a "vortex finder," and passes out as "overflow" through the top of the cone.

Advantages:

Centrifugal forces separate particles without the need for large settling tanks

Good at removing solids with diameters of approximately 30 microns and larger

Limitations:

During upsets in flow or pressure drop, the rotary motion in the cone may be interrupted,

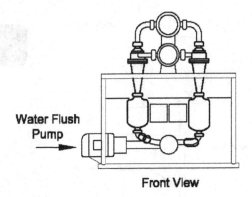

Front View

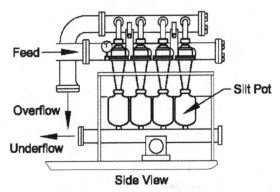

Side View

FIGURE 2-8B Schematic of a eight-cone hydrocyclone assembly.

possibly causing solids to carry over into the overflow liquid

Experience wear problems; some manufacturers offer replaceable lines to handle wear problems

Large pressure drops

Limited ability to handle surges in flow

Rarely used as the only solids' removal device, although they can remove high loadings of solids, making them very useful as a first step in solids removal.

If filters are used as a second step:

> Hydrocyclone can greatly lengthen the filters' cycle time

> Filters will provide removal of the smaller-diameter solids and protect against carryover from the hydrocyclone during upsets

Ability of a hydrocyclone to separate a certain diameter solid particle (fineness of separation) is affected by the:

> Differential pressure between the inlet and overflow

> Density difference between the solid particles and the liquid

> Geometry and size of the cone and inlet nozzle

Pressure drop through the cone is the critical variable in:

> Terms of affecting fineness of separation

> Itself a function of flow rate

The lower the flow rate the:

> Lower the pressure drop

> Coarser the separation

Typically, hydrocyclones are operated with 25 to 50 psi (140 to 275 kPa) pressure drop.

Many theoretical and empirical equations have been proposed for calculating fineness of separation. All reduce to the following form for a hydrocyclone of fixed proportions:

$$d_{50} = K \left[\frac{D^3 \mu}{Q_w \Delta SG} \right] \tag{2-16}$$

where

D = major diameter of hydrocyclone cone

D_{50} = solid particle diameter that is recovered 50% to the overflow and 50% to the underflow, microns (μ)

μ = slurry viscosity, cp (Pas)

Q_w = slurry flow rate, bpd (m^3/hr)

ΔSG = difference in specific gravity between the solid and the liquid

K = proportionally and shape constant

The diameter of solid particle that is recovered 1 to 3% to the overflow and 97 to 99% to the underflow is

$$d_{99} = 2.2d_{50} \qquad (2\text{-}17)$$

The flow rate through a hydrocyclone of fixed proportions handling a specified slurry is given by:

$$Q_w = K'(\Delta P)^{1/2} \qquad (2\text{-}18)$$

where

Q_w = flow rate, bwpd (m³/hr)

K' = proportionally and shape constant

ΔP = pressure drop, psi (kPa)

Equations 2-16, 2-17, and 2-18 can be used to approximate the performance of a hydrocyclone for different flow conditions, if its performance is know for a specific set of flow conditions.

Solids discharge in the underflow slurry is performed in either an:

Open system

Closed system

In an open system:

Slurry is rejected through an adjustable orifice at the apex of the cone to an open trough.

Orifice can be adjusted to regulate the flow rate of the water leaving with the solids.

It can allow oxygen into the system.

In a closed system:

A small vessel called a "silt pot" is connected to the apex, which remains open.

A valve is located at the bottom of the silt pot and is usually closed.

Solids pass through the apex and collect in the bottom of the silt pot.

The valve at the bottom of the silt pot is opened periodically to reject the solids.

Opening or closing can be manual or automatic.

Hydrocyclones may be put on line individually, thus providing some ability to account for changes in flow rate.

When specifying a hydrocyclone, the following information must be provided:

> Total water flow rate
>
> Particle size to be removed and the percentage of removal required
>
> Concentration, particle size distribution, and specific gravity of particles in the feed
>
> Design working pressure of the hydrocyclone
>
> Minimum pressure drop available for the hydrocyclone

With the above information the designer can select equipment from various manufacturers' catalog descriptions.

Centrifuges

Centrifuges are used to separate low-gravity solids or very high percentages of high-gravity solids.

Centrifugal force is used to rapidly separate solids from the liquid.

They require extensive maintenance and can handle only small liquid flow rates, therefore are not commonly used in water treating applications.

Flotation Units

Primarily used for removing suspended hydrocarbons from water.

They make it possible to remove small particles in the liquid stream.

Principles of operation:

> Gas is dispersed into the water and forms bubbles approximately 30 to 120 microns (μ) in diameter.
>
> Bubbles form on the surfaces of the suspended particles, creating particles whose average density is less than that of water.
>
> These rise to the surface and are mechanically skimmed.
>
> Chemicals called "float aids" are normally added in the feed stream.
>
>> Aid in coagulation of solids and attachment of gas bubbles to the solids.

Optimum concentration and chemical formulation of float aids are normally determined from batch tests in small-scale plastic flotation models on site.

Because of the difficulty of predicting particle removal efficiency with this method, it is not normally used to remove solids from water in production facilities.

Disposal Cartridge Filters

Disposal cartridge filters are simple and relatively lightweight.

They can be used to meet a variety of filtration requirements.

A typical cartridge filter is shown in Figure 2-9

Water enters the top section and must flow through one of the filter elements to exit through the lower section of the vessel.

Top head of the vessel is bolted so that the cartridges can be changed when the pressure drop across them reaches an upper limit.

A PSV can be installed in the vessel to prevent excessive differential pressure between the upper and lower sections of the vessel.

They are available in a wide variety of materials, and they provide a range of performance options.

Cartridges are available with manufacturers' particle size ratings from 0.25 microns (μ) and above.

When selecting a filter cartridge, the designer must determine what the manufacturer's rating actually means in terms of removal percentage.

Solids removal and allowable flow rates:

Vary greatly from manufacturer to manufacturer, even if the cartridges are made of the same material

Make it difficult to develop generalized relationships between the water flow rate and filter area

Require relying on manufacturers' information when selecting and sizing a cartridge filter system.

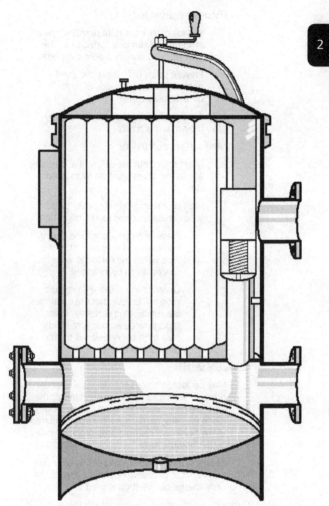

FIGURE 2-9 Cartridge filter vessel. Courtesy of Perry Equipment Corp.

When designing a water treatment system that includes cartridge filters:

It is desirable to select a fixed-pore filter medium and absolute rated filters.

Fixed-pore cartridges:

Provide more consistent particle removal efficiencies from one cartridge to the next than do nonfixed-pore cartridges

Prevent solids unloading and media migration during periods of high differential pressure

Are usually given absolute ratings by their manufacturers

Nonfixed-pore cartridges:

May be used but the differential pressure across the filters must be monitored closely

High differential pressure may cause solids unloading and media migration

Pressure drop through the filter will decrease and may be below the limit when the cartridge is scheduled to be changed.

Operator checking the pressure drop will believe that the cartridges are functioning correctly, even though large amounts of solids may have been released to the downstream water.

Solids unloading:

May be avoided by using a high differential pressure switch to continuously monitor the pressure drop or by changing the cartridges when the pressure drop is still small compared to the maximum pressure drop recommended by the manufacturer.

Resulting frequent changing of the cartridges may result in excessive operating costs if the early charge-out method is used.

Cartridge filters typically have low solids-loading limits.

Cartridges can only absorb a relatively small amount of solids before they must be changed.

Cartridges to improve solids loading:

Pleated construction of a thin filter medium increases the effective filter surface area.

Increased surface area provides for higher flow rates and solids-loading capacities than a cylindrical cartridge of the same medium.

Multilayered design of media are characterized by the following:

Provides in-depth filtration

Have progressively smaller pores as the water moves from the outside to the inside of the cartridge

As pore size changes, particles are trapped at different depths within the filter, allowing higher solids loadings but typically decreasing flow rates slightly.

Cartridge filters have low solids-loadings.

It is common to install primary solids removal equipment upstream of the cartridge filters.

Typical systems include either a hydrocyclone or a sand filter followed by the cartridge filter.

Upstream equipment removes the larger solids and reduces the amount of solids that the cartridges must remove, therefore extending the time between cartridge changes.

Spare filter vessel are characterized by the following:

Allows changing cartridges without reducing water flow rates

Most common system arrangements include:

Three 50% vessels

Four 33% vessels

Number of vessels selected depends on a cost analysis and on operating preference

Other factors to consider in the selection of cartridge filters:

Type of filter medium and its characteristics

Polypropylene cartridges are a better selection than cotton for water service, since cotton swells.

Compatibility of filter membranes and binders with chemical additives or impurities in the water should be checked

When specifying a cartridge filter the following information should be included:

Maximum water flow rate

Particle size to be removed by filtration and the presence of removal required

Solids concentrations in the inlet water

Design working pressure of the filter vessel

Maximum pressure drop available for filtration

Backwashable Cartridge Filters

Available in a variety of designs using:

Metal screens

Permeable ceramic

Consolidated sand

Advantages:

Simple and lightweight

Backwashable

Media used filters provide filtration of particles between 10 and 75 microns (μ)

Differential pressure considerations:

Have low solids-loading limits, therefore they have potentially short intervals between backwash cycles.

Do not expose filters to differential pressures over (170 kPa) as the particles may become too deeply embedded in the pores to be removed by backwashing.

With proper maintenance and repeated backwashing, this type of filter may last up to two years.

Regeneration or "Backwashing"

Involves flowing clean water through the filter in the opposite direction of the normal filtration

Requires an acid backwash as well

Solids trapped in the filter media are then forced out of the filter and carried away with the backwash fluid

Process is quicker and may be less costly than changing cartridges

Flow rate of fluid required for backwash is specified by the manufacturer

Disadvantages:

> Filtered water must be stored and then pumped through the filter.
>
> Resulting backwash fluid must then be directed to another storage medium.
>
> A method and equipment for disposing of the backwash fluid, which can be contaminated with oil or acid used in the backwash cycle, must also be provided.

It is available in a variety of designs including:

Cartridge filter vessel (shown in Figure 2-9)

Manifold design

> Each cartridge is housed in a separate housing and the housings are manifolded.
>
> It is possible to backwash individual filters while the other filters continue to operate normally.

When specifying a backwashable cartridge filter, the designer should include the following:

Maximum water flow rate

Particle size to be removed by filtration and the percentage of removal required

Solids concentration in the inlet water

Design working pressure of the filter vessel

Maximum pressure drop available for filtration

Designer should contact manufacturer for detailed information on selecting filters of this type.

Granular Media Filters

Granular media filter and *sand filter* refer to a number of filter designs in which fluid passes through a bed of granular medium.

Filters consist of a pressure vessel filed with the filter media (refer to Figure 2-10).

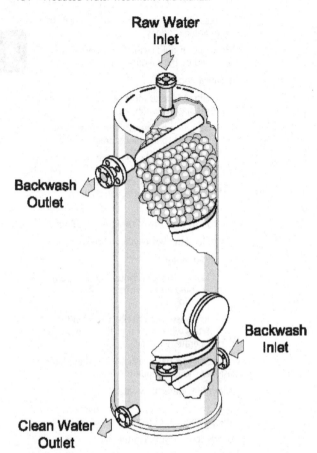

Raw Water Inlet

Backwash Outlet

Backwash Inlet

Clean Water Outlet

FIGURE 2-10 Downflow granular media filter. Courtesy of CE Natco.

Media support screens prevent the media solids from leaving the filter vessel.

Water to be filtered may flow either down (downflow) or up (upflow) through the media.

As the water passes through the media, the small solids are trapped in the small pores between the media particles.

Downflow filters:

Designed as either

"Conventional" (Figure 2-11)

Designed for an approximate flow rate range of 1 to 8 gpm/ft.^{2} (2.5 to 20 m^3/hr m^2)

"High flow rate" (Figure 2-12)

May have flow rates as high as 20 gpm/ft. (249 m^3/hr m^2)

At higher velocities, a deeper penetration of the bed is achieved, allowing a higher solids loading (weight of solids trapped per cubic foot of bed)

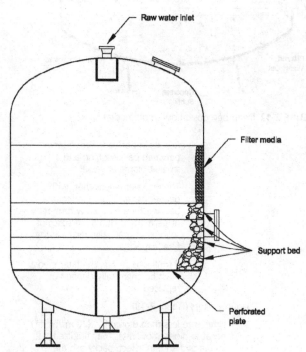

FIGURE 2-11 Conventional graded bed filter.

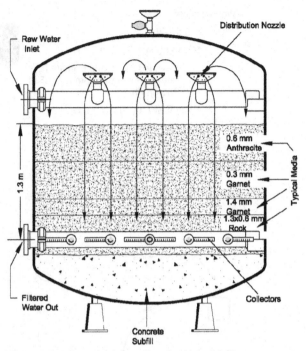

FIGURE 2-12 Deep bed downflow (multimedia) filter.

Results in both a longer interval between backwashing and a smaller-diameter vessel

The disadvantage is that, with deeper penetration, inadequate backwashing may allow formation of permanent clumps of solids that gradually decrease the filter capacity

If fouling is severe, the filter media must be chemically cleaned or replaced

Upflow filters (Figure 2-13):

Limited to less than 8 gpm/ft.^{2} (20 m^3/hr m^2) because higher flow rates may fluidize the media bed and, in effect, backwash the media.

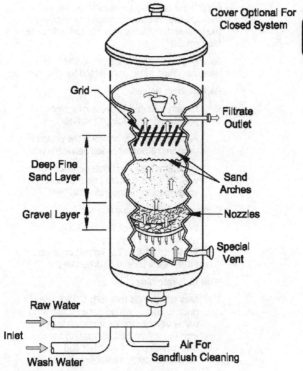

FIGURE 2-13 Deep bed upflow filter.

Backwashing

Granular media filters must be cleaned periodically by backwashing to remove filter solids.

Process involves fluidizing the bed to eliminate the small pore spaces in which solids were trapped during filtration.

The small solids are then removed with the backwash fluid through a media screen that prevents loss of media solids.

Media may be fluidized by flowing water upward through the filter at a high rate or by introducing the water through a nozzle that produces high velocities and turbulence within the filter vessel.

Recycle pumps may be used to pump water through the fluidization nozzle to decrease the total water volume required to fluidize the filter media.

As with backwashable cartridge filters, the backwash fluid must be collected for disposal.

Backwashing process

> Usually initiated because of a high-pressure drop through the filter
>
> Regular schedule, provided the pressure drop limit is not exceeded between cycles
>
> Cycle time for a sand filter depends on the water's solids content and the allowable solids loading of the individual filter

Conventional downflow filters

> With flow rates of less than 8 gpm/ft.^{2} (20 m^3/hr m^2) typically can remove ½ to 1½ lb/ft.^{2} (2.4 to 7.3 kg/m^2) of solids of filter media before backwashing.

High flow-rate filters

> May remove up to 4 lb/ft.^{2} (19.5 kg/m^2) prior to backwashing because the high water velocity forces small solids farther into the media bed, increasing the effective depth of the filter and thus the number of pores available to trap solids.

Upflow filters

> May remove up to 6 lb/ft.^{2} (29.3 kg/m^2) because the upward flow loosens and partially fluidizes the bed, allowing greater penetration by the small solids.

Upflow versus downflow selection:

Governed by the influent suspended solids content and the preferred time between backwash cycles

> Downflow filters are used when the suspended solids content of the influent is below 50 mg/l.
>
> Upflow filters are used for a suspended solids content range of 50 to 500 mg/l.

Table 2-3 provides a comparison of typical influent flow rates and solids loadings.

Table 2-3 Typical Parameters for Granular Bed Filters

Type	Flow Rate (m³/hr m²)	(gpm/ft²)	Solids Loading* (kg/m²)	(lb/ft²)
Conventional downflow	2.4–19.6	1–8	2.4–7.3	0.5–1.5
High-rate downflow	19.6–48.9	8–20	7.3–19.5	1.5–4
Upflow	14.7–29.3	6–12	19.5–48.8	4–10

*Weight of solids trapped per unit area of media prior to backwashing.

Granular media filters fall in the category of nonfixed-pore filters because the filter media are not held rigidly in place.

If not backwashed promptly, granular media filters can unload previously filtered solids.

Media migration, however, is usually not a problem because media screens are usually built into the filter vessel, preventing the media from leaving the vessel.

Granular media filters use:

Sand

Gravel

Anthracite

Graphite

Pecan or walnut shells

Bed may be made of a single material or of several layers of different material.

Increases the solids loading by forcing the water through progressively smaller pores (large-to-small)

Effects of variable pore size distribution:

Variable pore size, depends on the random distribution of the media solids after backwashing

Due to variable pore size, cannot be given an absolute rating

Consistently remove 95% of all 10-micron and larger solids.

Backwash flow rates are characterized by the following:

Vary with specific filter designs

Specified by the manufacturer

Some designs require an initial air or gas scour [10 to 15 psig (69 to 103 kPa) supply] to fluidize the bed

Especially true for filters handling produced waters that contain suspended hydrocarbons that can coat the filter media

Several cycles of scour followed by flushing may be required during the backwash operation

Detergents may also be needed to aid in cleaning the filter media

Raw water is usually used for backwash

When the backwash cycle is complete, water is allowed to flow through the filter for a period of time until the effluent quality stabilizes; only then is the filter put back on stream

A filter's ability to trap particles smaller than the pore is greatly aided by the addition of polyelectrolites and filter aids.

Chemicals promote coagulation in the line leaning to the filter and aid the formation of a chemical or ionic bond between these small particles and the filter medium.

A specific filter may be capable of removing 90% of the 10 μ and larger particles without chemicals and 98% of the 2 μ and larger particles with chemicals.

Applications:

Commonly used as the first filtration step (normally called "primary filtration") prior to cartridge filters (known as "secondary filtration")

System works well because:

Granular media filter removes the bulk of the large solids, thus increasing the cycle time for replacing cartridges.

Cartridge filters:

Remove the small solids to the required size

Catch any solids released by the sand filter due to unloading

2

Tables 2-4 and 2-5 provide typical operating and design parameters for two types of granular media filters.

Specific manufacturers should be contacted to select a standard granular media filter and obtain detailed sizing and operating information.

Table 2-4 Typical Operating and Design Parameters for a Specific Upflow Filter

A. Operating Parameters

Service rate	14.6 to 29.3 m^3/hr m^2 (6 to 12 gpm/ft.^{2})
Chemical treatment	Polyelectrolytes at 0.5 to 5 ppm
	Determine if needed by bench tests
Flush rate	Temperature-dependent (34.2 to 48.9 m^3/hr m^2, or 14 to 20 gpm/ft.^{2})

Regeneration time sequence
Cycle 1:

Drain	2 to 5 minutes (drain water to top of sand bed)
Fluidize bed	5 minutes with air or natural gas
Flush	10 to 20 minutes (until water is clear)

Cycle 2:

Drain	3 to 5 minutes (drain water to top of sand bed)
Fluidize bed	5 minutes with air or natural gas
Flush	10 to 20 minutes (until water is clear)
Settle	5 minutes
Prefilter	15 to 20 minutes depending on water quality

B. Design Parameters

Service rate	14.6 to 29.3 m^3/hr m^2 (6 to 12 gpm/ft.^{2}) of filter area
Inlet solids	Will hold up to 49 kg of solids m^2 (10 lb/ft.^{2}) of filter area (400 ppm maximum)
Inlet oil	Up to 50 ppm
Total outlet solids	2 to 5 ppm without chemical treatment
	1 to 2 ppm with chemical treatment
Outlet oil	Less than 1 ppm
Cycle length	2-day minimum
Fluidize gas flow	55 to 90 m^3/hr per m^2 (3 to 5 cfm/ft.^{2}) surface area (supply pressure of 83 to 109 kPa (12 to 15 psig))
Freeboard area	50 to 70% of total media depth
Bed expansion	Approximately 30% during flush cycle
Particle size removal	By theory, can be calculated from smallest sand (Barkman and Davidson)

C. Miscellaneous Data

1. If inlet water contains above 15 ppm oil, a solvent or surfactant wash may be required during regeneration cycle number 1.
2. Sizing of media
   1st layer: 32 to 38 mm gravel, 101 mm thick (1¼ to 1½ in. gravel, 4 in. thick)
   2nd layer: 10 to 16 mm gravel, 254 mm thick (⅜ to ⅝ in. gravel, 10 in. to 60 in. thick)
   3rd layer: 2 to 3 mm sand, 305 mm thick (2 to 3 mm sand, 12 in. thick)
   4th layer: 1 to 2 mm sand, 1524 mm thick (1 to 2 mm sand, 60 in. thick)

Table 2-5 Typical Operating and Design Parameters for a Specific Downflow Filter

A. Operating Parameters

Service rate	11.0 m^3/hr m^2 (4.5 gpm/ft.^{2})
Chemical treatment	20 ppm blend of cationic polyelectrolyte and sodium laminate
Regeneration	
Backwash	4 minutes at 41.6 m^3/hr m^2 (17 gpm/ft.^{2})
Rinse	4 minutes at 11.0 m^3/hr m^2 (4.5 gpm/ft.^{2})

B. Design Parameters

Service rate	4.9 m^3/hr m^3 (2 gpm/ft.^{2})
Inlet solids	< 20 ppm
Inlet oil	< 10 ppm

C. Miscellaneous Data

1. Sizing of media

	Thickness		Size		Specific
Kind	(mm)	(in.)	(mm)	(in.)	Gravity
Anthracite (top)	457	18	1.0 to 1.1	—	1.5
Sand	229	9	0.45 to 0.55	—	2.6
Garnet	76	3	0.2 to 0.3	—	4.2
Garnet	76	3	1.0 to 2.0	—	4.2
Gravel	76	3	4.8 × No. 10	3/16 × No. 10 Mesh	2.6
			Mesh		
Gravel	76	3	9.5 × 4.8	⅜ × 3/16	2.6
Gravel	76	3	19.0 × 9.5	¾ × ⅜	2.6
Gravel	76	3	38.1 × 19.0	1½ × ¾	2.6
Rock	76	3	50.8 × 38.1	2 × 1½	2.6

2. Filter may need detergent in backwash.
3. High amounts of oil during upset conditions may necessitate solvent washing the filter media.
4. Media can stick together and form balls with excessive chemicals or oil in the inlet and may require the bed to be replaced or cleaned.
5. Backwash rates in excess of 41.6 m^3/hr m^3 (17 gpm/ft.^{2}) may cause carryover of anthracite, especially when backwash water is cold.

To select a granular media filter, the designer should specify the following:

Maximum water flow rate

Particle size to be removed by filtration and the percentage of removal required

Solids concentration in the inlet water

Maximum pressure drop available for filtration

Diatomaceous Earth Filters

Diatomaceous earth filters are used for filtration of 0.5 to 1.0 μ particles.

They were used in the past to remove very fine solids because they were the least costly filters available in this range.

Manufacturers have developed cartridge filters that can effectively remove 0.25 μ solids.

DE filters remove solids by forcing the water through a filter cake of diatomaceous earth.

The filter cake is built up on thin wire screens of corrosive-resistant materials such as stainless steel, Monel, or Inconel.

A large number of wire screens, called "leaves," may be arranged within the vessel to provide a large surface area for filtration.

Typical flow rate through DE filter screens ranges from 0.5 to 1 gpm/ft.^{2} (1.2 to 2.4 m^3/hr m^2).

A DE filter is shown in Figures 2-14a and 2-14b.

This process involves precoating the leaves with a thin layer of DE, which is introduced as a slurry (refer to Figures 2-15a and 2-15b).

After the precoat, the water is introduced and filtration begins.

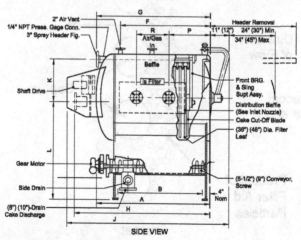

FIGURE 2-14A DE filter. Courtesy of US Filter Corp.

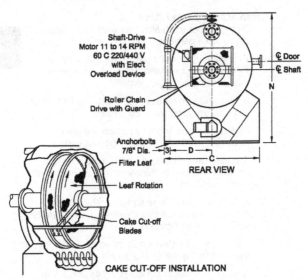

FIGURE 2-14B DE filter. Courtesy of US Filter Corp.

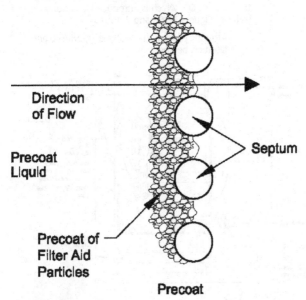

FIGURE 2-15A Principles of DE filtration. Courtesy of Johns Manville Corp.

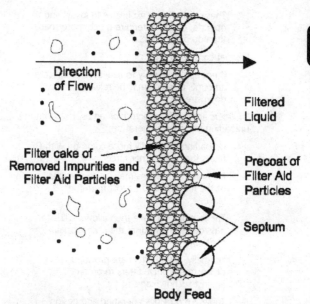

FIGURE 2-15B Principles of DE filtration. Courtesy of Johns Manville Corp.

A filter aid, such as DE and cellulose fiber must be mixed with the water to promote an even build-up of filter cake and maintain the filter cake's permeability.

This combination is called "body feed."

The weight of the body feed should be roughly equal to the weight of the solids to be filtered.

When the pressure drop reaches the high limit, usually between 25 and 35 psig (170 and 240 kPa), the filter cake must be backwashed from the leaves and the process started over with the precoat.

DE filters:

Require slurry mixing tanks

Require injection pumps

Require large quantities of body feed in addition to the filter vessel itself

These systems are expensive to install and to operate, and they require much more space than do other filters.

Precoat considerations

If not applied evenly to all leaves, significant amounts of solids may be released downstream.

DE filters are nonfixed-pore type and suffer from unloading and media migration.

Unloading is typically more common with DE filters than with other filters.

Pressure fluctuations may cause portions of the filter cake to be lost from the leaves.

The loss of filter cake then allows solids to pass downstream until the cake is again built up.

Normally, guard filters are provided downstream of DE filters to protect against such leakage.

Table 2-6 provides operating and design parameters for a typical DE filter.

Chemical Scavenging Equipment

Chemical scavenging equipment typically requires:

Chemical storage facilities

Mix tanks

Injection pumps

Depending on the injection rate:

Storage facilities may simply be drum racks or

Storage may be a small atmospheric tank

Premixed chemical scavengers may also be purchased in drums or bulk tanks if quantities used are relatively small.

To select the best method of storage and mixing and to size injection pumps, it is necessary to calculate the chemical injection rate.

The following method provides an estimate of chemical usage based on the reaction stoichiometry.

Table 2-6 Operating and Design Parameters for a Typical DE Filter

A. Operating Parameters

Service rate	1.2 to 2.4 m³/hr m² (0.5 to 1 gpm/ft.²)
DE bodyfeed	2 to 5 ppm DE/ppm suspended solids
Regeneration time sequence	
Drain	1 to 5 minutes
Sluice	5 minutes
Fill and add precoat	3 minutes
Circulate	5 to 15 minutes
DE precoat:	
Amount	0.5 to 1.0 kg/m² (10 to 20 lb/100 ft.²)
Filter slurry	30 to 60% water
Circulate rate	2.4 to 4.9 m³/hr m² (1 to 2 gpm/ft.²) (4.5 fps)

B. Design Parameters

Service rate	1.2 m³/hr m² (0.5 gpm/ft.²)
Inlet solids	< 20 ppm
Inlet oil	< 10 ppm
Total outlet solids	< I ppm
Regenerate at 20-psig pressure drop across filter	

C. Miscellaneous Data

Wet bulk density DE	240 to 320 kg DE/m³ (15 to 20 lb DE/ft.³)
Dry bulk density DE	112 to 240 kg DE/m³ (7 to 15 lb DE/ft.³)
Specific gravity DE	2.3
Cycle length	2 to 3 days
Screen material	Polypropylene, plain weave, 33 × 42 count 630 denier warp, twist direction 3.5 Z, weight 201 g/m² (5.92 oz./yd²), heat set for permeability of 730 m³/hr m² (40 cfm/ft.²) and scoured. Stainless steel can be used.

Note: Bulk density of Perlite filter aid is one-half that of DE. When Perlite is used, the above guidelines should be adjusted on an equivalent volume.

Specific chemical suppliers and equipment manufacturers should be contacted to assist in making final equipment selections.

The required injection rate of a chemical scavenger may be calculated as follows:

Field units

$$W_{cs} = 1.09 \times 10^{-5} Q_w SG_w CO_2 RMW_{cs} \qquad (2\text{-}19a)$$

SI units

$$W_{cs} = 7.5 \times 10^{-4} Q_w SG_w CO_2 RMW_{cs} \qquad (2\text{-}19b)$$

where

$$W_{cs} = \text{mass flow rate of chemical scavenger,}$$
$$\text{lb/day (kg/day)}$$

$$Q_w = \text{water flow rate, bwpd (m}^3\text{/hr)}$$

$$SG_w = \text{water specific gravity}$$

$$CO_2 = \text{inlet oxygen concentration in water,}$$
$$\text{ppm}$$

$$R = \text{stoichiometric reaction ratio between}$$
$$\text{the scavenger and oxygen lb/hr O}_2$$
$$\text{(kg/mol/hr O}_2\text{)}$$

$$MW_{cs} = \text{chemical scavenger molecular weight,}$$
$$\text{lb/mol}$$

Equation 2-19 indicates the mass flow rate of the chemical scavenger's active ingredient.

The pump injection rate depends on the concentration of active ingredient in the mixed chemical solution.

Chemical manufacturers can assist in determining the best solution concentration and the resulting volumetric injection rate.

The required injection rate of catalyst may be calculated as follows:

Field units

$$W_c = 7.7 \times 10^{-7} Q_w SG_w C_{ca} \tag{2-20a}$$

SI units

$$W_c = 5.3 \times 10^{-2} Q_w SG_w C_{ca} \tag{2-20b}$$

where

$$W_c = \text{mass flow rate of catalyst (COCl}_2\text{), lb/day}$$
$$\text{(kg/day)}$$

$$C_{ca} = \text{catalyst concentration, ppm (normally}$$
$$C_{ca} = 0.001 \text{ ppm)}$$

The volumetric injection rate depends on the mixed solution concentration of the catalyst.

Manufacturers may be able to provide a premixed solution of scavenger and catalyst. This solution should be considered because it will decrease the amount of storage, mixing, and injection equipment required.

DESIGN EXAMPLE: SOLID REMOVAL PROCESS

Complete Water Injection System

Given:

Injection rate = 1300 gpm

Injection pressure = 770 psig

Filtration required to = 2 micron

Subsurface water data = 0 ppm oil, 0 ppm O_2, 80 ppm sand

Distribution = 97% greater than 35 microns, remainder between 35 and 2 microns

Water rate available = 1500 gpm

Water pressure at surface = 60 psig

Lift gas flow rate = 6 MMSCFD

Step 1. Select subsurface water as the source.

Proceed to Step 8 in the procedure.

Step 8. Lift gas is used to produce the water.

Proceed to Step 9.

Step 9. Size a two-phase separator to separate the lift gas from the water.

Step 10. Filtration to 2 microns is required.

Step 11. Solids concentration over 35 microns.

$(0.97) \times (80 \text{ ppm}) = 77.6 \text{ ppm}$

This is greater than 25 ppm; therefore, proceed to Step 12.

Step 12. Select a hydrocyclone and proceed to Step 13.

Step 13. Specify a hydrocyclone unit.

The particle size to be removed is 35 microns. The water flow rate is 1300 gpm.

The minimum pressure drop available for the hydrocyclone must be calculated. If the drop through the system can be limited to the 60 psig available, then filter charge pumps will not be required.

Water level within the inlet separator can be maintained with a valve controlling the rate at which the excess water is dumped overboard. Flow through the system will be controlled by the injection pump suction rates; therefore, no allowance for a hydrocyclone inlet flow control valve is made.

Filtration requirements dictate the use of cartridge filters to meet the 2 micron requirements. Just prior to replacing the filter elements, these filters will require a maximum 20 psi pressure drop.

The pumps will require a small positive pressure at their suction to prevent cavitation problems. An allowance of 5 psig should be sufficient.

Piping losses can be neglected because the cartridge filters will normally operate far below the 20 psi allowed.

System pressure drops can be summarized as follows:

Inlet pressure = 60 psig

Control valves = 0 psi

Cartridge filters = 20 psi

Piping losses = 0 psi

Pressure at pumps = 5 psi

Minimum pressure drop available for the hydrocyclone = 35 psi

Therefore, a hydrocyclone should be selected based on the following specifications:

Water flow rate = 1300 gpm

Solids removal to = 35 microns

Pressure drop = 35 psi

Removal to 35 microns is not acceptable; therefore, proceed to Step 18.

Step 18. Filtration to 2 microns is required. Select replaceable cartridge filters and proceed to Step 19.

Step 19. Select a standard cartridge filter unit to meet the following specifications.

Water flow rate = 1300 gpm

Solids removal to = 2 microns

Solids content of water = 2.4 ppm

Operating pressure = 25 psig

Pressure drop available = 20 psi

Step 20. No cartridge filter is required as a guard filter.

Step 21. There is no oxygen in the water, and the water mixes readily with the produced water without forming any precipitates; therefore no chemical injection is required.

NOMENCLATURE

A = Cross-sectional area of the particle, ft.^{2} (m^2)

Cca = Catalyst concentration, ppm

CD = Drag coefficient

CO_2 = Inlet oxygen concentration in water, ppm

D = Vessel's internal diameter, in. (m)

Dm = Particle diameter, μ

$d50$ = Particle diameter that is recovered 50% to the overflow and 50% to underflow

$d99$ = Particle diameter that is recovered 1% to the overflow and 99% to the underflow

F = Factor that accounts for turbulence and short-circuiting

FD = Drag force, lb (kg)

G = Gravitational constant, 32.2 ft./s^2 (9.81 m/s^2)

H = Height of water, ft. (m)

Hw = Height of the vessel water, in. (m)

Hw = Water height, in. (m)

K = Proportionality and shape constant for solids particle removal

K = Proportionality and shape constant for flow rate vs. pressure drop

L_{eff} = Effective length in which separation occurs, ft. (m)

L_{ss} = Seam-to-seam length, ft. (m)

$MWCS$ = Chemical scavenger molecular weight, lb mol/hr (kg mol/hr)

P = Operating pressure, psia (kPa)

Q = Volumetric vapor flow rate at pop and T, actual ft.^{3}/min (m^3/hr)

Qg = Gas flow rate, MMSCFD (std m^3/hr)

Qw = Water flow rate, bwpd (m^3/hr)

R = Stoichiometric reaction ratio between the scavenger and oxygen, lb mol/hr O_2(kg mol/hr O2)

Re = Reynolds number

SGw = Water specific gravity

T = operating temperature, °R (K)

$(tr)w$ = water retention time, min

Vt	= terminal settling velocity of the particle, ft./s (m/s)
W	= width, ft. (m)
Wc	= mass flow rate of catalyst (COC_{12}), lb/day (kg/day)
WCS	= mass flow rate of chemical scavenger, lb/day (kg/day)
Aw	= fractional cross-sectional area of water
Γw	= fractional water height within the vessel (hw/d)
Γ	= height to width ratio, (Hw/W)
ΔP	= pressure drop, psi (kPa)
ΔSG	= difference in specific gravity of the particle and the water
μw	= water viscosity, cp (Pas)
ρ	= density of the continuous phase, lb/ft.^{3} (kg/m^3)

Appendix A
Definition of Key Water Treating Terms

Introduction

Most of these terms were defined in Parts 1 and 2 when initially introduced.

This appendix is not intended to be a comprehensive listing of all terms.

Produced Water

The well stream from the reservoir typically contains varying quantities of water, commonly referred to as "produced water."

Produced water source can be from

1. An aquifer layer underlying the oil and or natural gas zones

2. Connate water found within the reservoir formation sand matrix

3. Water vapor condensing from the gas phase as the result of Joule-Thompson expansion/cooling effects occurring from pressure reduction up the well bore and across wellhead chokes

4. Water-bearing formations not directly in communication with the hydrocarbon reservoir

5. Or a combination of the above

Produced water is typically salty and contains varying quantities of:

Dissolved inorganic compounds and salts

Suspended scales and other particles

Dissolved gases

Dissolved and dispersed liquid hydrocarbons

Various organic compounds

Bacteria

Toxicants

Trace quantities of naturally occurring radioactive materials

Other miscellaneous sources of water from within the processing facilities (e.g., from drains, glycol regeneration units, etc.) are sometimes mixed with produced water for treatment and disposal.

Regulatory Definitions

The terminology for "total oil and grease," "dispersed oil," and "dissolved oil" may vary with location and specific test standard used by the authorities having jurisdiction.

These terms should be applied with caution and should conform to the regulations and test standards applicable to the specific location.

Oil Removal Efficiency

Produced water treating equipment performance is commonly described in terms of its "oil removal efficiency."

This efficiency considers only the removal of dispersed oil and neglects the dissolved oil content.

DOI: 10.1016/B978-1-85617-984-3.00003-1

For example, if the equipment removes half of the dispersed oil contained in the influent produced water, it is said to have a 50% oil removal efficiency.

For a specific piece of equipment or an overall system, the oil removal efficiency can be calculated using the following equation:

$$E = [1 - (C_o/C_i)] \times 100, \qquad \text{(A-1)}$$

where

E = oil removal efficiency, %

C_o = dispersed oil concentration in the water outlet (effluent) stream, ppm (mg/l)

C_i = dispersed oil concentration in the water inlet (influent) stream, ppm (mg/l)

The performance can be described by determining the inlet and outlet oil concentrations and the associated oil droplet size distributions at the equipment inlet and outlet.

This information can then be used to define the oil removal efficiency for any given oil droplet size or range of droplet sizes. This concept is further discussed in Part 1.

Total Oil and Grease

"Total oil and grease" is defined as the combination of both the dispersed and dissolved liquid hydrocarbons and other organic compounds (i.e., "dissolved oil" plus "dispersed oil") contained in produced water.

This term is referenced in certain regulatory standards and is commonly used to evaluate water treating system design.

Total oil and grease consists of normal paraffinic, asphaltic, and aromatic hydrocarbon compounds plus specialty compounds from treating chemicals. The measurement of total oil and grease is dependent on the analysis method used.

Dispersed Oil

Produced water contains hydrocarbons in the form of dispersed oil droplets, which, under proper conditions, can be coalesced into a continuous hydrocarbon liquid phase and then separated from the aqueous phase using various separation devices.

The diameters of these oil droplets can range from over 200 microns to less than 0.5 microns and may be surrounded by a film (emulsifier) that impedes coalescence.

The relative distribution of droplet sizes is an important design parameter and is influenced by the hydrocarbon properties, temperature, down-hole operating conditions, presence of trace chemical contaminants, upstream processing and pipe fittings, control valves, pumps, and other equipment that act to create turbulence and shearing action.

These oil droplets are collectively defined as "dispersed oil."

Conventional water treating systems commonly used by the oil and gas industry remove only the "dispersed oil."

This text focuses on the design of water treating systems that remove and recover "dispersed oil."

Dissolved Oil

Produced water contains hydrocarbons and other organic compounds that have dissolved within the aqueous phase and cannot be recovered by conventional water treating systems.

Fatty acids are likely to be present within paraffinic oils and naphthenic acids within asphaltic oils.

These organic acids, aromatic components, polar compounds (also called nonhydrocarbon organics), and certain treating chemicals are slightly soluble in water and collectively make up the organic compounds found in solution of the aqueous phase.

The portion of these components that are dissolved into the produced water is defined as "dissolved oil."

Dissolved oil is microscopically indistinguishable within the aqueous phase since it is solution at the molecular level and cannot be separated from the produced water by means of coalescence and/or gravity separation devices.

Treatment methods for removal of dissolved oil are not covered in this text.

However, the oil and gas industry is currently evaluating treatment methods such as biotreatment, air stripping, adsorption filtration, and membranes.

These designs are typically prototypical in nature and require a larger capital investment, a greater maintenance work effort, and more space, and may result in by-products having disposal problems more onerous than those associated with the disposal of dissolved oil.

The confined space typically available on an offshore platform presents a real challenge in developing a suitable water treating process for removing dissolved oil from produced water.

Dissolved Solids

Several inorganic compounds are soluble in water.

The total measure of these compounds found in solution with produced water is referred to as "total dissolved solids" (TDS).

When these compounds are found in solution with the produced water, they are referred to as "dissolved solids."

The most common water-soluble compound in produced water is sodium chloride.

A number of other compounds collectively comprise the dissolved solids contained in produced water. These are discussed in Part 1.

Suspended Solids

Produced water and oil contain very small particulate solid matter held in suspension in the liquid phase by surface tension and electrostatic forces.

This solid matter is referred to as a "suspended solid" and may consist of small particles of sand, clay, precipitated salts and flakes of scale, and products of corrosion such as iron oxide and iron carbonate.

When suspended solids are measured by weight or volume, the composite measurement is referred to as the "total suspended solids" (TSS) content.

Part 1 provides a detailed discussion on suspended solids.

Scale

Under certain conditions, the dissolved solids precipitate or crystallize from the produced water to form solid deposits in pipe and equipment.

These solid deposits are referred to as "scale." The most common scales include calcium carbonate, calcium sulphate, barium sulphate, strontium sulphate, and iron sulfide. Scale is further discussed in Part 1.

Emulsion

An "emulsion" is an oil and water mixture that has been subjected to shearing, resulting in the division of oil and water phases into small droplets.

Most emulsions encountered in the oil field are water droplets in an oil continuous phase and are referred to as "normal emulsions."

Oil droplets in a water continuous phase are referred to as "reverse emulsions."

Appendix B

Water Sampling Techniques

Sampling Considerations

Any water analysis method is only as good as the "sample" used to represent the effluent stream.

Sampling of a continuously flowing stream containing two or more phases (e.g., oil and water) is difficult unless the mixture is completely emulsified or is a very fine stable dispersion.

Since the sampling techniques for oil concentration measurement and particle size distribution differ in some aspects, they are described separately here.

Sample Gathering for Oil Concentration Measurement

Generally, the larger the sample the more likely it is to be representative. However, for practical reasons, the sample size varies from 15 ml to about 1 l.

Typically, the smaller samples are used for daily analysis, whereas the larger samples are used for monthly regulatory compliance purposes.

The smaller the residual oil droplets, the more evenly dispersed they are likely to be.

Care should be exercised to avoid sampling the surface of a liquid (since this is not truly representative). "Isokinetic" (which means equal linear velocity) sampling in midstream is the best, but is rarely possible.

The sample probe must be inserted so that the velocity profile remains undisturbed, thereby getting a realistic particle distribution and, thus, a higher accuracy.

The general guidelines are:

Flush the sample line thoroughly and take the sample quickly.

Sample after a pump or a similar turbulent area where the stream is well-mixed.

Obtain the sample from a liquid-full vertical pipe, if possible.

Sample bottles should be scrupulously clean and preferably made of glass. Oil or other organic material can adhere to the walls of a plastic container and give erroneous readings.

Never use a metal container or a metal cap. The water can corrode it and become contaminated with corrosion products.

Bottles used at oily sites or handled by an operator with oily hands can have thin surface films, and washing can leave detergent residue, both of which can give rise to erroneous and high oil readings.

General guidelines one should follow to improve measurement accuracy are as follows:

Use only glass or inert plastic (e.g., Teflon) stoppers. Cork or other absorbent materials must not be used unless covered with aluminum foil.

DOI: 10.1016/B978-1-85617-984-3.00004-3

Do not rinse or overflow the bottle with the sample because an oil film will appear on the bottle and give a false reading.

Cap the sample and prepare a label immediately with an indelible, smear-proof marking pen. Attach it to the bottle immediately.

Analyze the entire sample and wash the bottle with solvent.

The person taking the sample must be well-trained and experienced and be able to recognize a spoiled or unrepresentative sample.

Samples must be correctly labeled immediately after being taken and any abnormal circumstances must be noted on the sample.

If any doubt exists, the sample should be discarded and a new one taken in a fresh container.

The sampling frequency depends on the practicality of sampling at each site or may also be specified by the authorities having jurisdiction.

A manned installation would require a higher-analysis frequency than an unmanned site, which may be less accessible.

For example, in the United States, the EPA states,

The sample type shall be a 24-hour composite consisting of the arithmetic average of results of 4 grab samples taken over a 24-hour period.
If only one sample is taken for any one month, it must meet both the daily and monthly limits. Samples shall be collected prior to the addition of any seawater to the produced water waste stream.

Sample Storage for Oil Concentration Measurement

If possible, perform the sample analysis as soon as the sample is obtained. If immediate analysis is

not possible as with certain samples (normally for regulatory reporting) that are sent onshore for analysis, then acidify the sample to pH 2 using hydrochloric acid (HCl) to preserve the sample against bacterial action and/or dissolve the precipitated calcium carbonate, which could cause difficulties separating the solvent phase from the water.

Acidification causes a higher total oil and grease concentration.

This is because acid reacts with organic salts to liberate organic acids, which are then extracted into the solvent.

This gives a higher reading in the analysis.

Therefore, when a sample has been acidified, the solvent extract should pass through a silica-gel column or similar material to remove these polar substances (organic acids).

In the case of analysis done immediately (for daily measurement), the sample should be acidified only if the approved analysis procedure requires it.

Sample Gathering for Particle Size Analysis

The following points are applicable when obtaining a sample for particle size analysis:

Flush the sample line thoroughly and take the sample quickly.

Obtain the sample from a liquid-full vertical pipe, if possible.

Use only glass or inert plastic (e.g., Teflon) stoppers. Cork or other absorbent materials must not be used unless covered with aluminum foil.

Do not rinse or overflow the bottle with the sample as this can put an oil film on the bottle and give a false reading.

Cap the sample and prepare a label immediately with an indelible, smear-proof marking pen. Attach it to the bottle immediately.

In addition to the above factors, other important factors that need to be considered when measuring particle size distribution are:

Avoid shearing of oil droplets across a sample valve.

Typically, when sampling produced water the sample valve is never fully open, since full flow is so intense that sampling may be almost impossible.

Consequently, the flow is restricted by controlling the valve partially open.

In this situation, the produced water is subjected to choking from high pressure down to atmospheric conditions.

The shearing within the sampling valve causes the oil droplets in the sample to break up into smaller droplets.

Avoid a "dissolved gas flotation effect."

The produced water sample is depressurized as it passes through the choke valve, and gas is liberated as minute gas bubbles.

These bubbles may coalesce with the oil droplets so the oil droplets adhere to the gas bubbles and rise to the sample surface.

This phenomenon is called "dissolved gas flotation." It will not alter the total concentration of oil in the sample, but could split the oil present into two separate fractions:

Dispersed oil, remaining in the water and not affected by the gas bubbles

Free oil, formed as a thin film on the water surface caused by the flotation

The free oil formed may easily adhere to the walls of the sample container as a thin and almost invisible film.

This film is easily lost as the initial sample is split into subsamples. The fraction of oil in the sample ending up as free oil may be as high as 50 to 80%.

One way of avoiding droplet shear and gas flotation effects is to conduct an online sample measurement.

This technique, however, requires specialized equipment, such as the Malvern Mastersizer, and is constrained in terms of the maximum pressure it can tolerate.

Alternatively, a sample pressure cylinder (bomb) can be used to avoid droplet shear and the gas flotation effect.

A sample in a stainless steel cylinder with a needle valve and a ball valve minimizes the shearing of droplets during sampling.

Avoid coalescence of oil droplets by stabilizing the sample.

A sample used for droplet size measurement may need to be stabilized to avoid coalescence of the small oil droplets to larger ones.

The propensity of a droplet to coalesce increases as the oil concentration of the sample increases.

This stabilization becomes more important when measuring samples with a high oil concentration.

Stabilization may be achieved by diluting the sample with a known amount of water.

This reduces the chance of droplet coalescence and also stabilizes the sample by reducing the salt concentration of the sample.

Further stabilization of the sample may be achieved by the addition of a viscous polymer solution and/ or surfactant (e.g., 2% sodium dodecyl sulphate).

However, one should be very cautious when adding such chemicals since the wrong surfactant could actually promote coalescence.

Sample may contain solids and other non-oil particles in addition to the oil droplets.

Produced water samples often contain solids and other non-oil particles in addition to oil droplets.

To determine only the oil droplet size distribution, one must first determine the size distribution for all particles within the sample and then determine the size distribution of particles left behind in the sample after solvent extraction.

Solvent extraction removes all of the oil droplets from the sample, leaving behind only the solids and nonextractable non-oil particles.

One can then block out those size ranges (corresponding to the solids and non-oil particles) from the initial size distribution to obtain a more representative size distribution.

Appendix C

Oil Concentration Analysis Techniques

Introduction

Several analytical techniques measure the amount of oil and grease in water.

These techniques may be broadly classified as either gravimetric or infrared (IR) absorbance methods and are described in detail here.

These methods are based on the extraction of oil and grease into a solvent. A sample may contain suspended solids, which have to be filtered.

> In this case, the sample must first undergo solvent extraction followed by filtration of the extract.

Several different solvents have been used. These include the following:

> petroleum ether
>
> diethyl ether
>
> chloroform
>
> carbon tetrachloride

Of these, petroleum ether and diethyl ether are highly flammable, whereas chloroform (although a very good solvent) and carbon tetrachloride are toxic.

Thus, these solvents are not recommended for use. Currently, 1,1,2-trichloro, trifluoroethane (Freon 113) is used when infrared (IR) absorbance is used for analysis.

However, these solvents are being phased out because of potential interference with the ozone layer in the atmosphere.

Studies are currently under way to find a replacement solvent. Potential candidates include

> hexane
>
> cyclohexane
>
> methylene chloride
>
> perchloroethylene
>
> a commercial hydrochlorofluorocarbon (DuPont 123)

When the gravimetric technique is used for analysis, 1,1,1-trichloroethane or dichloroethylene may also be used.

"Total oil and grease" is defined by the measurement procedure stipulated by the authorities having jurisdiction. Important variations that could give different results for the same sample are:

Number of extractions performed on the water sample. Multiple extractions with Freon on the same water sample will generally give a higher oil concentration than a single extraction.

The solvent-to-sample ratio. A higher solvent-to-sample ratio will also give a higher oil concentration.

Determination of IR absorbance at multiple wavelengths. This variation will give a higher oil concentration as opposed to absorbance measurement at a single wavelength.

Use of silica gel. This variation is discussed below.

DOI: 10.1016/B978-1-85617-984-3.00005-5

Determination of Dissolved Oil and Grease

The dissolved oil and grease content is first determined by measuring the total oil and grease content and then subtracting the measured dispersed oil and grease content.

The measured dispersed oil is obtained by removal of the dissolved oil and grease from the solvent with silica gel.

This can be expressed by the following equation:

dissolved oil and grease = total oil and grease − dispersed oil and grease

Dissolved organic matter is generally either polar or of low molecular weight.

To remove the dissolved organic matter, the solvent extract is contacted (either in an adsorption column or by intimate mixing) with activated silica gel (e.g., Florisil) or alumina.

The materials not adsorbed by the silica gel are described as "dispersed" oil and grease.

In the absence of silica gel, filtration can be used to remove the dispersed oil and grease.

In this technique the filtrate, obtained after passing through a 0.45-μ filter paper, is left behind on the filter paper.

The U.S. EPA uses the term "petroleum hydrocarbons" for dispersed oil and grease.

The choice of measuring total oil and grease versus dispersed oil and grease is important from an operational standpoint because conventional water treating technology can reduce the concentration of dispersed oil and grease, but not the concentration of dissolved oil and grease.

However, measurement techniques are strictly specified in some countries (e.g., the United States) and left open to negotiation or operator discretion in others.

Gravimetric Method

This method is the U.S. EPA-required method to measure oil and grease in produced water for regulatory compliance purposes in the United States.

A detailed procedure can be found in EPA 413.1. In this method the water sample is extracted with Freon 113 and the extract evaporated to remove the solvent.

The weight of the residue is related to the concentration of oil in the water sample.

Generally, for the same water sample, this method gives a lower value for oil and grease concentration than the IR absorbance method because of the loss of volatile organics during the evaporation process.

Advantages

Stipulated by the EPA as the technique to be used to measure oil in produced water discharge for regulatory purposes.

Method is simple and well understood.

Disadvantages

Requires samples be collected and preserved according to EPA protocol for shipment to an onshore laboratory.

Is time-consuming.

Uses Freon (a CFC) as a solvent.

Lower limit of measurement is 5 mg/l.

Not applicable to light hydrocarbons that volatilize below 70°C.

Infrared (IR) Absorbance Method

For infrared absorbance methods (e.g., EPA 413.2 and EPA 418.1), the water sample is extracted with Freon 113.

The IR absorbance of the extract is measured at single, or multiple, wavelength(s) to give the oil concentration.

In this method, the water sample is often acidified to prevent any salts from precipitating out (e.g., iron sulfide).

IR absorbance at multiple wavelengths results in a higher oil concentration measurement than if a single wavelength was used.

Advantages

Fast and convenient for offshore surveillance.

The lower limit of measurement is 0.2 mg/l.

Disadvantages

Uses Freon (a CFC).

The lower limit is 0.2 mg/l.

Analysis of Variance of Analytical Results

In 1975, the U.S. EPA had a set of six different samples analyzed by a number of laboratories using both the gravimetric and IR methods.

Scatter in reported analysis included sampling errors as well as analytical errors.

The true values were taken to be the average of the reported values (excluding those of extreme scatter) and are as follows:

(*Note*: Error is defined as the difference of an observation from the best obtainable estimate of the true value, which in this case is the arithmetic mean.)

It is interesting to note that the gravimetric value is lower than the infrared value of each of the samples.

This is expected since the solvent evaporation step in the gravimetric process causes some loss of volatile organics, leading to lower results than the IR method. Errors ranged from 0 to 241%.

The worst laboratory had errors between 49% and 98%, whereas the best laboratory had errors up to 8%.

Particle Size Analysis

The oil droplet size distribution is one of the key parameters influencing water treating equipment selection.

Therefore, accurate measurement of the oil droplet size distribution is an important task.

Sample	1	2	3	4	5	6
Gravimetric	201.7	209.3	95.5	77.6	33.7	19.1
Infrared	261.3	265.4	131.0	110.0	50.7	23.0

Another important parameter is quantifying the size distribution upstream and downstream of production equipment, such as control valves.

Oil and other particles in produced water range in size from less than 1 μ up to several hundred microns.

Although many particles found in produced water are not spherical, for practical purposes, the particles are represented by equivalent spheres.

Droplet Size Measurement Equipment

Three different types of equipment are commonly used for droplet size measurement.

Each has its advantages and disadvantages. First, establish the information desired before selecting the equipment type:

1. **Coulter counter.** The Coulter counter consists of two electrodes immersed in a beaker of sample water, which contains a sufficient number of dissolved ions to easily conduct an electrical current.

    The negative electrode is located inside a glass tube, which is sealed except for a tiny hole or orifice on the side of the tube.

    The positive electrode is located in the water sample beaker.

    A constant electrical current is passed from the positive electrode to the negative

electrode through the orifice.

When a nonconductive particle passes through the orifice, a change in electrical resistance occurs between the two electrodes, which is proportional to the particle volume.

A fixed volume of water containing suspended particles is forced through the orifice.

As each particle passes through the orifice, the increased resistance results in a voltage that is proportional to the particle volume.

The series of pulses produced by a series of particles passing through the orifice are electronically scaled and counted, yielding a particle size distribution.

One must realize that the particle "diameter" given by the counter is the diameter of a fictitious equivalent sphere with the same volume as the real particle.

This equipment has some limitations because the size range is limited. In addition, the samples have to be suspended in an electrolyte solution, which can prove difficult if the sample is totally soluble in the solution.

However, the Coulter counter does provide both frequency and volume distributions against volumetric particle size.

2. **Light (laser) scattering counters.** These include instruments that are based on the principle of light absorption/total scatter, or light blockage, to detect particles in a fluid.

   Water flows through a sensor cell, and as each particle passes through the intense beam of light in the sensor, light is scattered.

   The instrument measures the magnitude of each scattered light pulse, which is proportional to the surface area of the particle.

   The particle diameter determined by the instrument in this case is the diameter of a sphere with the same surface area as the particle.

   Laser diffraction systems are widely accepted due to their ease of use, wide size range, and simple sample preparation.

   However, as an optical technique, it is still subject to variations in response from particle shape and refractive index and is unable to give frequency information (that is, the number of particles within a given size range).

   However, these techniques can provide relative frequency information (that is, percent of total particle volume within a given size range).

3. **Microscopy.** In this technique, the droplet size distribution is determined by observing the water sample under a microscope and visually measuring the size of the droplets.

   Often a magnified photograph is also used for visual determination.

   This technique has the advantage of being able to distinguish between oil droplets and non-oil particles.

   A microscope also helps to see first hand if there are any extreme shape factors.

   However, the technique generally uses a very small sample volume and therefore may not be representative.

References

Kawahara, F.K (1991). A study to select a suitable replacement solvent for Freon 113 in the gravimetric determination of oil and grease. *EPA* (2 October).

Patton, C.C (1986). Water sampling and analysis. *Applied Water Technology*. Campbell Petroleum Series, Norman, OK.

Particle sizing: Past and present. *Particle Sizing Review, Filtration and Separation Journal*, (July/August).

API RP 45. (1981). Recommended practice for analysis of oil-field waters. Washington, DC: API.

INDEX

Note: Page numbers followed by 'f' indicate figures and 't' indicate tables.

Printed in the United States
By Bookmasters